VIROLOGY

An information profile

VIROLOGY

An information profile

ROBIN AND DAVID NICHOLAS

Mansell Publishing Limited

ISBN 0 7201 1673 2

Mansell Publishing Limited, 6 All Saints Street, London N1 9RL

First published 1983

© Robin Nicholas and David Nicholas 1983

Distributed in the United States and Canada by
The H. W. Wilson Company, 950 University Avenue, Bronx, New York 10452

All rights reserved. No part of this publication may be reproduced or transmitted in any form or by any means, electronic or mechanical, including photocopy, recording, or any information storage or retrieval system, without permission in writing from the publishers or their appointed agents.

British Library Cataloguing in Publication Data

Nicholas, Robin
   Virology, an information profile.
   1. Virology—Information services
   2. Virology—Bibliography
   I. Title   II. Nicholas, David, *1947–*
   576'.64   Z5185.U5
   ISBN 0-7201-1673-2

Printed and bound in Great Britain at
The Camelot Press Ltd, Southampton

Contents

Preface — vii

PART I OVERVIEW OF VIROLOGY AND ITS LITERATURE — 1

1 History and scope of virology — 3
  1.1 History and development of the subject — 3
  1.2 A definition — 11
  1.3 Relationship of virology to other disciplines — 12
  1.4 Virology as a profession — 14
  1.5 The information needs of virologists — 18

2 Organizations and their role in virology — 21
  2.1 Introduction — 22
  2.2 Finding out about organizations — 23
  2.3 Organizations in virology — 26
  2.4 International organizations — 28
  2.5 Commercial companies — 39
  2.6 National organizations — UK — 41
  2.7 National organizations — USA — 51
  2.8 National organizations — other countries — 58

3 Conferences — 63
  3.1 The role of conferences — 63
  3.2 Guides to forthcoming conferences — 64
  3.3 The major virology conferences — 65

4 The literature of virology — 71
  4.1 Journals — 72
  4.2 Reviews — 87
  4.3 Books — 94
  4.4 Conference proceedings — 107
  4.5 Dissertations — 108
  4.6 Reference works — 108

5 Searching the literature — 115
  5.1 Guides to the literature — 115
  5.2 Tracing and locating journals — 117
  5.3 Tracing and locating periodical articles — 120
  5.4 Subject bibliographies — 137
  5.5 Tracing and locating virology books — 139

5.6	Tracing and locating conference proceedings	143
5.7	Tracing and locating dissertations	145
5.8	Classification schemes: finding virology documents in libraries and bibliographies	146
5.9	Epilogue	157

6 Culture collections 161
6.1 Importance of culture collections 161
6.2 Culture collections and patents 163
6.3 Culture collection catalogues 164

7 Legislation and laboratory safety 165
7.1 UK legislation and codes of practice 165
7.2 US legislation and safety rules 167

PART II   BIBLIOGRAPHY 169

PART III   DIRECTORY OF ORGANIZATIONS, CULTURE COLLECTIONS AND LIBRARIES 197

INDEX 225

Preface

Virology is arguably the most diffuse of all biological sciences with much of its subject matter still residing in the fields from which it so recently emerged — medicine, veterinary science, biology and agriculture. Indeed the responsibility for its welfare and development, certainly in the UK and USA — the prime centres of virus research — is not in its own hands; in both countries its affairs are in fact run by microbiological societies.

Logically enough this scatter of interest is reflected in the literature. Information channels for such an apparently small subject are numerous and dispersed over a wide front. Furthermore, many of these information channels are partly hidden from view, buried as they are under the subject labels of other disciplines.

It is the role of this guide to provide the structure and coherence that are so patently missing from the literature, furnishing as it were a one-place reference to all sources of information on virology regardless of the discipline from which they originate.

Inevitably, it is a guide to the academic/research literature because, after all, this represents the bulk of all communication within the field; however, the increasingly important industrial/business literature has been covered in full.

The guide has been conceived as a self-help tool and as such it addresses itself to those with little formal knowledge of information sources. Indeed it is our express intention to make information more accessible and, we hope, more palatable, to this very audience, who anyway must form the bulk of virologists.

Whilst no stones have been left unturned in the quest for information sources it has not been our wish to itemize every possible source, no matter how esoteric or marginal. To do this we felt would be to betray the very purpose of the work — to pinpoint the *key* sources. Without the aid of selection and criticism, information seekers would only find themselves confronted by another information jungle — that of information on information. Accessibility has also been an important factor in the choice of information sources — there is surely little point in knowing that a particular source exists if: (a) the chances of obtaining it are zero; or (b) the mechanics of its location is likely to be unacceptably complex and time-consuming.

Acknowledgements

Thanks go to: Kay and Chris Nicholas for typing the manuscript and providing valuable editorial comment; Mr D. E. Gray and Mrs M. E. Davidge, Librarians at the Central Veterinary Laboratory (Weybridge), for advice and provision of many valuable references; the World Health Organization for their cooperation; the Institute of Biology for permission to include results of surveys published in *The Biologist*; Geoff Wood for his many useful suggestions; and finally Rita Nicholas for her continuing inspiration.

Part I

OVERVIEW OF VIROLOGY AND ITS LITERATURE

1 History and scope of virology

1.1	History and development of the subject	3
1.2	A definition	11
1.3	Relationship of virology to other disciplines	12
1.4	Virology as a profession	14
1.5	The information needs of virologists	18

1.1 HISTORY AND DEVELOPMENT OF THE SUBJECT

Dating the birth of any subject is usually a somewhat arbitrary and speculative practice. Commonly the subject's origins are established by pinpointing the date when the subject's founding father performed the crucial experiment or made the salient observation. Thus immunology's origins are traced to Jenner (1798), bacteriology's to Pasteur (1860) and genetics' to Mendel (1865). In the case of virology its roots can be traced to the activities of two figures, one of whom carried out the crucial experiment (Ivanovsky) and the other who made the fundamental observation (Beijerinck).

In 1892 Ivanovsky, a Russian, published a paper in the *Bulletin of the Academy of Imperial Sciences, St Petersburg* describing how a disease of the tobacco plant could be transmitted by means of the sap from infected plants after it had been passed through a filter capable of retaining bacteria and other particles of a size beyond the lowest limits of optical microscopy. Rather than believe that the agent of tobacco mosaic disease was an extremely small but conventional bacterial-like organism, as was originally thought, Martinius Beijerinck, a Dutchman, declared in 1898 that it was one of a unique group of organisms and coined the term *virus* — a word then treated as a synonym for poison. Beijerinck's view was held at that time to be tantamount to heresy; unfortunately for Beijerinck confirmation of his belief was not to be available in his lifetime.

Tobacco mosaic virus (TMV) is, of course, a plant virus, as distinct from an animal virus. Animal virology itself dates from 1897 when Loeffler and Frosch, publishing in *Zentralblatt für Bakteriologie, Parasitenkunde, Infektionskrankheiten und*

Hygiene, described the filter-passing property of the virus of foot-and-mouth disease. Very shortly afterwards a number of other animal disease viruses were identified including African horse sickness and myxomatosis. Then in 1902 Reed and Carrol of the United States Army Commission, established to investigate yellow fever, announced in *American Medicine* that this disease was also due to a filterable virus — the first pathogen of man to be included in the list of filterable viruses. Rabies, against which Pasteur had attempted vaccination without identifying the causative agent, was first described as a virus in the *Annales de l'Institut Pasteur, Paris* later that year.

By the time Roux published what turned out to be virology's first review article in the *Bulletin de l'Institut Pasteur* in 1903 a total of nine filterable viruses were known.

Table 1.1 Milestones in the history of virology: a chronology

1892	Ivanovsky establishes the existence of plant viruses
1897	Loeffler and Frosch establish the existence of animal viruses (in their particular case foot-and-mouth)
1898	The term *virus* coined by Beijerinck
1902	Reed and Carrol establish the existence of the first virus to infect humans (yellow fever)
1903	Virology's first review article published by Roux in *Bulletin de l'Institut Pasteur*
1915–1917	Twort and d'Hérelle establish the existence of bacteriophages
1921	d'Hérelle publishes the first major textbook in virology, *The bacteriophage*
1922	One of the first major research programmes mounted by the Medical Research Council (UK) into filterable diseases
1930	First major conference dealing comprehensively with virology held in Paris, the International Congress for Microbiology
1939	Virology's first journal published, *Archiv für die Gesamte Virusforschung*
1939–1940	Thanks to the introduction of the electron microscope the first direct visualization of a bacteriophage and virus
1947	The World Health Organization begins its research programme into virus diseases
1950s	The 'golden age of phage research' begins
1953	*Advances in Virus Research* (virology's own annual review) first published
1955	The field's core journal — *Virology* — first published
1967	The first abstracting service devoted to virology — *Virology Abstracts* — first published
1968	The First International Conference for Virology
1977	Smallpox eradicated

Whilst the list gradually expanded during the next decade, detailed information on the basic properties of viruses was still lacking, primarily because viruses could still be identified only by the damage they inflicted on their hosts. The search for more knowledge received a boost during the period 1915–1917 when the results of a series of important experiments, conducted independently by F. W. Twort and F. d'Hérelle, demonstrated a virus capable of infecting bacteria. A notion that perhaps bacteriophages ('bacteria-eaters'), as they became known, could be used to combat bacterial infections in humans and animals was soon proved unfounded. However the real value of bacteriophages was that they offered to the researcher considerable experimental and practical advantages as compared with the use of animals and plants. These advantages were quickly exploited by research workers who, using well-established bacteriological techniques, reported many of the fundamental properties of 'phages' in the following years. Much of this work was published in 1921 in the classic work *The bacteriophage* by d'Hérelle, which represented the first major textbook in virology. A few years later *Filterable viruses*, edited by Rivers (1928), summarized the 'state of the art' and covered the known viruses of the period.

The journals that provided the mouthpiece for much of the early virological research were on the whole established medical publications, usually with a bias towards pathology. Examples of such journals include the *British Journal of Experimental Pathology*, the *American Journal of Pathology*, the *Journal of the American Medical Association*, the *Journal of Experimental Medicine*, *The Lancet* and those titles previously mentioned. The authors publishing in these journals represented many countries, mainly the USA, Germany, France and the UK, and worked in such establishments as: the Rockefeller Institute for Medical Research, where much of the polio and early tumour virus work was carried out; the Kaiser Wilhelm, later to become the Max Planck Institutes; the Instituts Pasteur in Paris and Brussels; the Institut für Infektionskrankheiten in Berlin, notable chiefly for its foot-and-mouth disease work; and the National Institute for Medical Research and the other Medical Research Council Laboratories in the UK, where a research programme into the filterable viruses, particularly influenza, was initiated as early as 1922. In the animal health field, the Royal Veterinary College in London deserves special mention for M'Fadyean's work on African horse sickness at the turn of the century. M'Fadyean, incidentally, was instrumental in the founding of the *Journal of Comparative Pathology and Therapy*, another important research organ, in 1888.

Viral research in universities worldwide was slow to make a significant contribution to the development of the field. One of the first entrants was the University of Manchester, where pioneering research into the culture *in vitro* of vaccinia virus was conducted in the late 1920s by Maitland, the then Professor of Bacteriology. His department is now known as the Department of Bacteriology and Virology. Concurrently with the developments at Manchester, a significant step forward was made in the USA with the appointment of d'Hérelle to the chair of Protobiology at Yale University. Another university quite early on the scene was Cambridge. Here important research was undertaken by a number of departments; the Departments of Biochemistry, Genetics, Pathology, Veterinary Medicine and Botany — nicely illustrating the diverse background from which virology emerged.

Forums for the discussion of viruses and viral disease in the early years were organized and their proceedings published by the various national medical and veterinary associations such as the Association of Veterinary Medicine (founded in 1863), which was responsible for the Rinderpest Congresses in the late nineteenth century. In the USA the interests of microbiologists were quickly recognized by the formation of the American Society for Microbiology (ASM) in 1899. Formerly known as the Society of American Bacteriologists, the ASM organized regular meetings to discuss rapidly growing areas.

A unification of the medical, veterinary and plant microbiological interests, which had previously moved in separate but parallel paths, came at the 1st International Congress for Microbiology in 1930 held at the Institut Pasteur in Paris. Its proceedings were published by the newly formed International Society for Microbiology, an offshoot of the longer-established International Union of Biological Sciences, which was founded in 1919. D'Hérelle and Bordet — another exponent of 'bacteriophagy' — were prominent speakers. Other subjects discussed were polio, herpes and vaccine viruses. The 2nd International Congress for Microbiology took place four years later in London. One entire section was given over to 'Viruses and virus diseases in animals and plants', some of the subjects covered being foot-and-mouth disease; cultivation of plant viruses; colds and influenza. Additionally, there was some consideration of the viral aetiology of tumours. Rous, who had discovered a transmissible fowl tumour virus in 1911, was present, as were C. Andrewes and W. M. Stanley, who were both to figure prominently in the next phase in virology's development.

Until the early 1930s the major advances in virology were made by those more interested in the effects caused by viruses in animals and, to some extent, plants than in the agent itself. This had led to some very important achievements such as the growth of herpesviruses in mice in 1929 and influenza virus in ferrets in 1931, which enabled their use as laboratory models for human infection. However, the emphasis was shifting as developments in other fields were being applied to the study of viruses. Stanley in 1935 developed relatively sophisticated biochemical techniques to provide samples of highly purified TMV, enabling partial analysis. A few years later, in 1937, these techniques facilitated the determination of the ribonucleoprotein composition of the virus by Bawden and Pirie. During this period a new and powerful tool, the electron microscope, was introduced into the field, permitting in 1939 and 1940 the first direct visualizations of bacteriophages and TMV. In 1939 *Naturwissenschaften* carried the first photographs taken by Kausche, Ruska and Pfankuch.

It was not just the techniques of the physical sciences that were being utilized; virology was now also interesting the chemists and physicists themselves. Many, like Delbrück, Böhr and Schrödinger, were attracted by the possibilities of unravelling the secrets of life itself, possibilities which they felt were inherent in this comparatively new field. The philosophies of this new school of thought were comprehensively propounded in a seminal publication known as 'Driemännerwerk' ('Green Paper'), published in *Nachrichten der Gesellschaft der Wissenschaften* in Göttingen in 1935. This contained the seeds of microbial genetics and gave an early impetus to molecular biology.

This shift in interest was naturally soon reflected in the journal literature, with medical journals losing their monopoly of the publication of viral research; the major beneficiaries were the journals of physics and biochemistry, in particular

Archives of Biochemistry, the *Journal of Biological Chemistry*, *Naturwissenschaften* and *Zeitschrift für Physik*. One of the first new books to embody the new approach was Doerr's *Handbuch der Virusforschung*, which was published in 1938. This was followed shortly afterwards by the publication in 1939 of the first journal devoted entirely to virology — *Archiv für die Gesamte Virusforschung* — which, surprisingly given its title, was published in English (and still is, as *Archives of Virology*). This event surely marked virology's coming of age as a fully fledged subject.

The Second World War, as elsewhere in science, proved a major setback for research in the field; in particular the dissolution of the Kaiser Wilhelm Institutes was a particular blow. Most of the staff were forced to emigrate to the USA, largely to the California Institute of Technology (Caltech) and the Cold Spring Harbor Laboratory at Long Island, New York. This one event effectively transferred the leadership in viral research from Germany to the USA.

At Caltech a definite decision was taken to concentrate resources and effort on the study of bacterial viruses belonging to the T-even (T_2, T_4 and T_6) *Escherichia coli* group. This group was to become the chosen model in many laboratories for exploring virus properties and for the examination of the gene itself. The subsequent discoveries had a profound effect on the nature and direction of post-war biology, not least those by Bernal and Fankuchen (1941), who showed that TMV was assembled from a large number of structurally identical protein subunits, and Avery (1944), who established DNA as the carrier of heredity. Although the latter discovery was made using bacteria, it gave a major boost to virology. To disseminate the information that was rapidly accumulating as a result of this intensive research programme, Delbrück at the Cold Spring Harbor Laboratory launched the first of what were to be annual phage courses, the proceedings of which have subsequently proved to be a major contribution to the literature of virology.

To some the spectacular advances that were being made in the experimental field of molecular biology were detracting from the importance of viruses as causes of disease. The founding of the World Health Organization (WHO) in 1947 restored the balance, to some extent, by its involvement with research and development in those virus diseases causing the greatest hardship throughout the world: smallpox, influenza and arthropod-borne diseases. The WHO established the World Influenza Centres at the outset and expanded these centres until in 1968 there were a total of 31 covering the major infectious diseases of man. Developments in animal virology followed shortly, with Coxsackie viruses being propagated in newborn mice in 1948 and, probably more importantly, the cultivation of poliomyelitis virus in cell culture by Enders (1949). Cell culture techniques had been developed independently for the study of embryology, genetics and cytology but had become increasingly applicable to the growth of animal viruses *in vitro*; i.e. providing a means of study without living host animals. The membranes and cavities of the embryonated domestic hen's egg were also used extensively for growing certain types of animal virus. The two techniques together provided the potential for growing large amounts of virus for vaccine production and enabled virus studies to be conducted on a more quantifiable basis.

Animal viruses were now rapidly becoming as amenable to detailed examination as phages. However, phages were still to play an important role over the next decade or so in our understanding of the inner working of the cell at the molecular level. In fact this period was to become known as the 'golden age of phage research' with important discoveries being made by: Hershey and Chase (1952), who established that the nucleic acid of phage enters the host cell and self-replicates; Zinder and Lederberg (1952), who found that the phage can incorporate part of the host gene and transfer it to another cell — a process known as *transduction*; and Geirer and Schramm (1956), who in their turn proved that pure nucleic acid from TMV was infective.

Not surprisingly this sequence of major research breakthroughs significantly boosted publishing activity in the virology field. Thus 1953 saw the launch of virology's own annual review entitled *Advances in Virus Research*, which was, and still is, published by Academic Press. In the ensuing years there was a veritable explosion in publishing, with the following journals appearing: *Virology* (1955), *Voprosy Virusologii* (1956), *Acta Virologica* (1957), *Progress in Medical Virology* (1958) and *Perspectives in Virology* (1959). Table 1.2 charts the growth of the journal/review literature.

Table 1.2 Growth of virology as a discipline as indicated by the number of journals and reviews in circulation

Year	Number of:	
	International virological journals	Regular review volumes
1941	1	0
1951	2	0
1961	5	4
1971	9	10
1981	13	11

While many great advances were being made on the research side, some attention was also being given to the practical aspects of veterinary and medical viruses and their vaccines. The first meeting of an association aimed at promoting the development and use of standardized techniques in microbiology was held in Lyon in 1955 by the International Association of Biological Standardization in cooperation with the WHO, its proceedings being published as *Developments in Biological Standardization* (1955–).

While it was true to say that in many respects virology had established itself as a viable subject field with its own body of knowledge and specialist methodology it still, nevertheless, lacked an organizational infrastructure and an institutional mouthpiece. Whilst the interests of American virologists had been looked after by the ASM and those of British virologists by the Society for General Microbiology (SGM), which was established in 1947, it was generally felt by the virology community that virology was not getting sufficient attention. As a direct result of this unease an important decision was taken at the 10th International Congress for Microbiology, held in Moscow in 1966 and organized by the IAMS, to set up a convening committee to prepare the ground for the 1st International Congress for Virology. This took place in Helsinki two years later and attracted 536 virologists, all of whom had made, in the eyes of the organizers, a significant contribution to virology. The proceedings of the meeting were published as *International virology I*. Simultaneously *Intervirology*, a monthly periodical of the virology section of the IAMS, was issued.

In the intervening period between the conception and birth of the International Congress an unprecedented expansion in the organization and literature of the subject took place. First came the field's very own abstracting service, *Virology Abstracts* (1967), a testament to the fact that the literature was larger than most virologists could keep abreast of. Then, from the two major organizations in the field — the ASM and the SGM — came two new journals, the *Journal of Virology* and the *Journal of General Virology* respectively. These major bodies had at last recognized the need for a separate outlet for papers devoted solely to virology.

At this time there was also a need to bring further unity to the various branches of virology to enable a greater concentration of resources and effort on the disease problems of man, domestic animals and plants. Such a realization was very much in the minds of the organizers of the 1st International Conference of Comparative Virology held in Quebec in 1969, and organized by the WHO in cooperation with the FAO, who had initiated the International Programme on Comparative Virology. Finally, in the interests of applied virology, the Pan-American Health Organization held the 1st International Conference on Vaccines against Viral and Rickettsial Diseases of Man in Washington in 1972.

But what of the research that was being reflected in a burgeoning literature (*Figure 1.1*)? As was seen earlier, phage research dominated the 1950s and first part of the 1960s, with the emphasis shifting from phage genetics to nucleic acid replication and regulation during this period. In medicine 'phage typing' established itself as a diagnostic tool for the routine identification of pathogenic bacteria. But then phage research for all intents and purposes went into decline, largely because in the real world phages are unimportant. As already mentioned, ideas for using them to combat bacterial disease were found very early on to be unrealistic. The small problems that phages posed in the dairy industry could never justify the financial commitment necessary to sustain research. But more importantly the techniques pioneered by phage workers had become readily available for the far more economically and medically important viruses of animals and man.

In the 1970s research into the role of viruses in cancer was virtually given a blank cheque, particularly in the USA, the powerhouse of virus research. The fundamental properties and behaviour of many candidate oncogenic viruses became explicitly known and even their genetic code sequenced as great advances in techniques followed.

Two previously neglected areas of virology that have recently attracted great interest are plant and invertebrate virology.

Viruses of plants rarely kill their host but simply multiply inside it, reducing the yield of vegetation or marring the appearance of flower crops. So widespread are virus diseases that until recently many plant varieties, potatoes in particular, were totally infected and had never been seen in a healthy virus-free condition.

Figure 1.1 Virological articles indexed by Biological Abstracts *between 1930 and 1980*

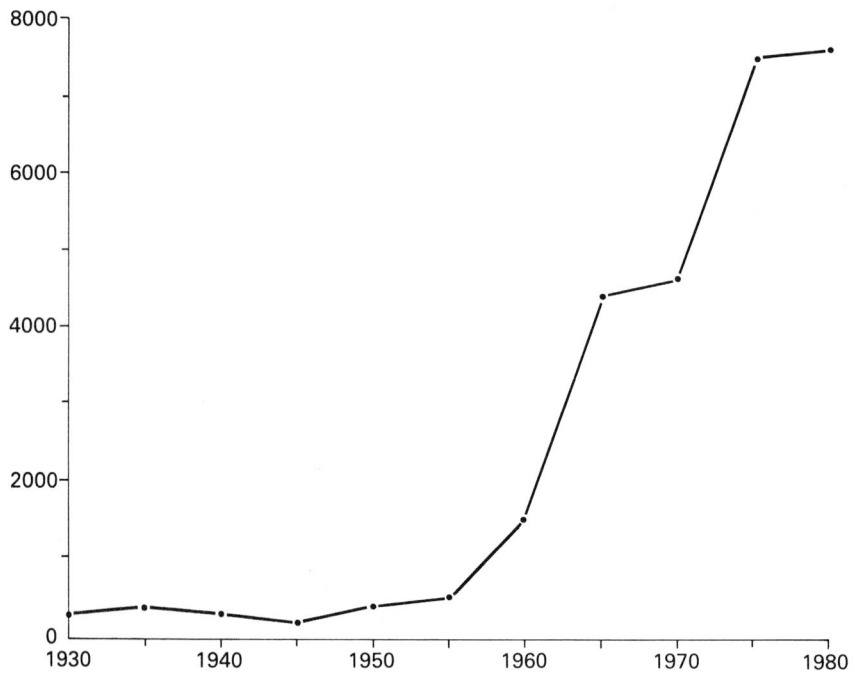

Research has been aimed at three main areas: first, the control of insects, which spread many viruses; secondly, the production of virus-free strains of plants by heat therapy treatment and apical meristem culture; and thirdly, in plant immunology by the isolation of resistance-associated protein — the plant equivalent of interferon. It is in this final area where most research is now taking place.

The viruses of invertebrates, like yellow fever, are combated most effectively by eliminating the carrier, with chemical insecticides. However the effect of chemical insecticides on the environment has led to an increasing interest in controlling pests biologically. It so happens that one group of pathogenic insect viruses, the Baculoviruses, seem to be confined to insects. Hopes that these 'viral pesticides' may be effective have been confirmed on a small scale in British forests and in tropical countries. The challenge to the virologist, here, is to provide the sensitive methods needed to monitor the genetic stability of the virus and to assess potential risk to the environment so that balance of risk and advantage can be weighed accurately. If this can be done the prospect for using these viruses to our species' advantage is exciting.

So where does the future of virology lie? Recent trends suggest that work in the field of immunology will play an increasing role in the prevention and the eradication of disease, as was witnessed for smallpox in 1977, largely as a result of successful vaccination therapy.

Other areas that will figure highly include the use of monoclonal antibodies for diagnosis and prevention of disease. More effective vaccines and antiviral agents such as interferon are also becoming available thanks to advances in genetic engineering and biotechnology.

Finally, from the point of view of pure virus research, although many human viral pathogens have been isolated and characterized, some are still being discovered such as the rotaviruses and the virus of African haemorrhagic fever. The nature of some, like the unconventional transmissible agents that cause the so-called 'slow virus infection', has yet to be defined, and others, such as viroids, the smallest known agents of infectious disease, may yet have great implication in certain animal and human infections.

1.2 A DEFINITION

It is usually convenient to divide the biological sciences into three groups according to the nature of their subject matter: *taxonomic, integrative* and *reductionist*. Taxonomic disciplines such as botany and zoology deal with a group of organisms having a common ancestry and historical development. Physiology, genetics and the like are concerned with the study of the common or specialized properties of living organisms and are therefore integrative. Reductive disciplines examine the elementary processes and functions of organisms at the molecular or atomic level, examples being biochemistry and biophysics.

Virology, however, does not slot easily into any of these groups because its subject matter, the viruses, cannot be defined by any of the standard criteria employed in animal and plant classifications. The much quoted 'A virus is a virus', attributable to Lwoff, whilst meaningless, testifies to the difficulty of explaining or defining viruses. The difficulty chiefly arises from the problem of reconciling their 'living' and 'non-living' properties. If we include the recently discovered 'viroids', viruses amount to the smallest biological entities capable of self-replication. They are most easily confused with bacteria — both being microorganisms capable of causing disease — but are distinguishable, amongst other ways, by containing only one type of nucleic acid, by their inability to multiply outside a living system, and by being unaffected by antibiotics.

The classification of the viruses themselves also leads to problems. The viral contribution to the fossil record is almost non-existent and that of bacteria is not extensive either. Viruses cannot therefore be grouped according to their evolutionary development. Because of this, bacterial classification and nomenclature has had to be based on an arbitrary selection of characteristics, and although some dissatisfaction is felt with this type of hierarchical, non-phylogenetic system it has generally become accepted by bacteriologists. *Bergey's Manual of determinative bacteriology* (1974), now in its eighth edition, has long been the definitive authority. Attempts at applying Bergey's system of latinized binomials to viruses proved unsatisfactory to virologists, largely because the criteria for classification were too heavily based on the effects the virus caused in its host rather than on the properties of the virus itself. In addition the nomenclature according to Bergey departed too far from the vernacular for the comfort of many virologists.

The majority of virus names derive from important clinical, pathological and epidemiological features of viral infections. Examples are contagious pustular dermatitis virus, a member of the pox group of viruses, and broad bean vascular wilt virus, a pathogen of plants. Viruses have also been named after the geographical location of their isolation (Coxsackie virus), and called after their discoverer (Rous and Epstein–Barr viruses). Some viruses are referred to only in their abbreviated or shortened form; thus reo corresponds to *r*espiratory *e*nteric *o*rphan virus; and arboviruses to *ar*thropod-*bo*rne viruses.

The most accepted and widespread method of classification is to group the viruses according to the type of host they infect:

Bacteria

Fungi

Plants

Invertebrates (chiefly insects)

Animals

Man

These may be further divided depending upon the level of interest. It should be noted that although we ourselves are animals we are often elevated to a separate group of our own because of the intense medical interest. This system is simple and convenient because, by and large, virologists tend to work with only one kind of host, although this is not necessarily the most logical division of labour. Some ambiguities nevertheless do result; thus plant viruses can multiply in insects and, more seriously, reoviruses can multiply in nearly all of the major groups of host.

While the above scheme is still used at a superficial and everyday level, the use of a rational taxonomic system based on principles of structure and molecular formation, as developed by the International Committee on Taxonomy of Viruses, has gained much acceptance. Indeed it has helped to establish virology as a self-sustaining discipline. *Figure 1.2* presents the essential details of the scheme.

1.3 RELATIONSHIP OF VIROLOGY TO OTHER DISCIPLINES

Historically, because of their small size, the detection of microbes was based purely on observation of abnormal changes produced in the animal or plant they infected, hence the early association with pathology: human, animal or plant pathology. As knowledge of microbes accumulated, the subject of microbiology developed to study their behaviour and characteristics. Their size placed restrictions on the methods that could be used to study them so techniques such as filtration, microscopy and sterilization developed. Equally their size and short generation times made them suitable for genetic studies and their role as infectious agents in animals enabled immunological methods to be applied to detect and diagnose the diseases they caused.

It was soon apparent however that bacteria, fungi and the 'filterable viruses' were quite distinct and the terms *bacteriology*, *mycology* and *virology* began to be used. The need of viruses for living cells in which to multiply led to an association with tissue culture, a once introverted branch of cytology. Tissue cultures, both of cells and of organs, have been used extensively for the study *in vitro* of viruses and their relationship with the cell, and have greatly facilitated vaccine production. The nature and extremely small size of virus particles — overlapping the range of

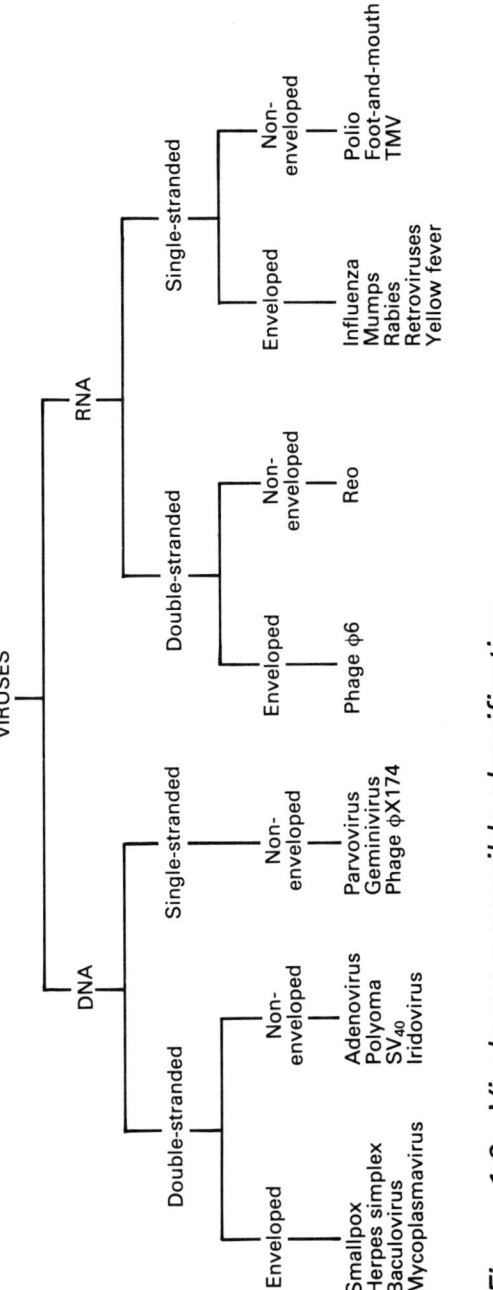

Figure 1.2 Virology: a possible classification

sizes of protein molecules and nucleic acid macromolecules — have meant that techniques could also be borrowed from biochemistry, and in turn, the methods of the physical sciences — chemistry and physics — were borrowed to help with the purification, quantification and characterization of viruses.

From the fusion of virology and biochemistry emerged a new area of biology, namely molecular biology, in which virology still plays an integral part. Here the emphasis is on the use of viruses as a tool in elucidating the basic regulatory mechanisms of cells at the molecular level.

A more recent development in virology has been in the study of cancer. A massive investment of resources and effort to implicate viruses as a cause of human cancer has led to the growth of a specialized area of virology with its own institutes and literature. Tumour virology, a hybrid of oncology and virology, consists chiefly of cell culture, molecular biology, immunology and epidemiology studies.

Finally the recent discovery of viral-like particules called *viroids* may lead to further branching of virology as techniques develop to study them.

Figure 1.3 attempts to portray graphically the web of relationships referred to in the above text.

1.4 VIROLOGY AS A PROFESSION

In the last few decades virology has emerged as a self-sustaining discipline with its own university departments, institutes and literature. There is, however, some evidence to suggest that virologists themselves have been slow to appreciate this fact. Furthermore, they do not always see themselves as virologists (particularly when confronted with a questionnaire from the authors of this book). 'Although I work with viruses I am really an ... immunologist/electron microscopist/biochemist/molecular biologist' is a very common reply. An examination of the membership lists of the Royal Society and Royal Society of Edinburgh reveals that only five of the 15 eminent virologists concerned described themselves *as* virologists or animal virologists; the rest miscellaneously saw themselves as biologists, microbiologists, pathologists, physicians, zoologists and one, curiously, as a protozoologist. Whilst this clearly demonstrates the interdisciplinary nature of virology it also shows a certain reluctance on the part of the people concerned to be labelled as a 'virologist'. Why should this be the case? Does virology's recent emergence account for the fact that there is, to the best of our knowledge, only one society (the little-known Society of Japanese Virologists) devoted to the well-being of virologists? Everywhere else virologists must share a stage dominated by bacteriologists. In fact it was only as recently as 1967 that a separate section was formed for virology in the International Association of Microbiological Societies. Immunology, also a relatively new subject and originally an applied branch of microbiology, now boasts a total of 35 national societies.

One possible explanation is that virology has been for the earlier and greater part of its history under too heavy an influence from the medical and, to a lesser extent, veterinary professions, whose members see themselves first and foremost as doctors and vets and not as virologists. Their virtual monopoly of many of the key research positions has prevented the rise of the science-educated virologists.

Another contributory factor may be that undergraduates do not identify early enough with virology because of the lack of first-degree courses in this subject. (This situation may be somewhat different in the USA where such courses do

Figure 1.3 Virology and its relationships with the major biological disciplines

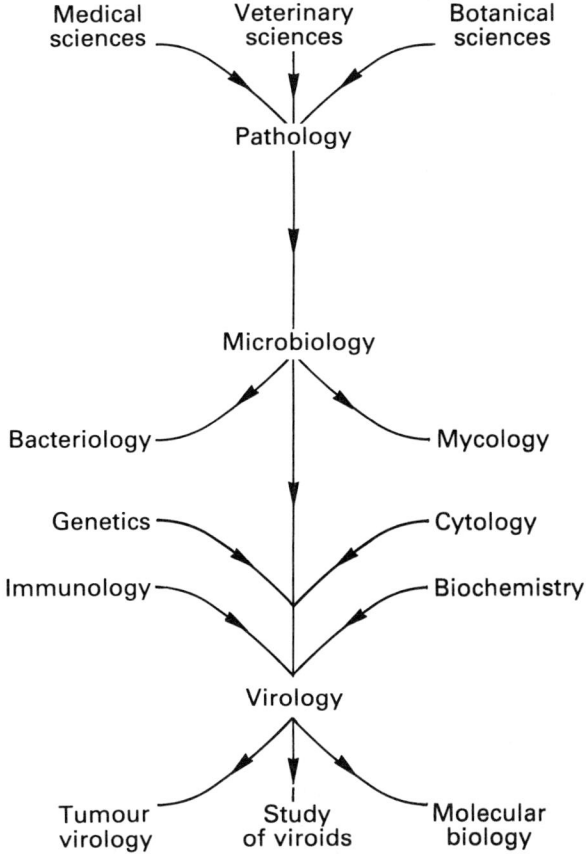

exist.) In the UK, virology is usually a component or, worse still, an option of microbiology. In most cases, therefore, microbiology graduates are unlikely to consider themselves, or be considered, virologists until they have obtained a higher degree or a post in a virology laboratory; usually both.

Interestingly the first destination of microbiology graduates has been the subject of a survey published by the Institute of Biology in *The Biologist* (1977). *Table 1.3* shows that just over a third found biologically related employment whilst about a quarter pursued higher qualifications. The same number that went into teaching were lost to biologically unrelated jobs.

Another, more recent, survey in *The Biologist* (1981), which examined the first destination of all biology graduates, showed that unemployment was again around 7%; this compared rather unfavourably with the corresponding figures for chemists and physicists, which were respectively 3% and 4%. Of those graduates in microbiology who found employment by far the biggest single destination was hospital laboratories, with the next largest number going to universities

Table 1.3   First destination of microbiology graduates in 1975

Higher degrees	24%
Teacher training	10%
Biologically related employment	36%
Unrelated employment	10%
Others and unknowns	13%
Believed unemployed	7%

Table 1.4   Employment distribution of microbiology graduates in 1979

Civil service	4%
Local government	7%
Hospitals	28%
Universities	19%
Chemical industry	14%
Other manufacturing industry	10%
Accountancy	5%
Others	13%

(*Table 1.4*). Accountancy is probably the most unusual and unexpected destination of biology graduates — perhaps it is a testament to their numeracy.

The same survey provides a picture of postgraduate employment for biologists (*Table 1.5*). About half of all postgraduates in 1979 took jobs in government or industry; one-fifth returned overseas, and about the same number continued studying, either with fellowships or in academic research. Unemployment is encouragingly lower for postgraduates.

Table 1.5   First destination of biology postgraduates in 1979

Research or academic study	6%
Teacher and other training	1%
Postdoctoral fellowship	12%
Employment	44%
Unemployment	3%
Not seeking employment (returning overseas)	21%
Unknown	14%

An alternative way of determining the employment situation for microbiologists/virologists is to examine the demand for their services as indicated by job advertisements for graduates appearing in the 'Situations Vacant' columns of major journals. Dent and Caldwell (1974) published two such surveys in *The Biologist* and distinguished between opportunities for graduate biologists and graduate technicians. Such a distinction may have been feasible some 20 years ago, but with the ever-increasing number of graduates entering the technical grades it is now rather meaningless. Therefore the two surveys have been fused. From *Table 1.6* it can be seen that government is the major employer of virologists, with most of the jobs being in the medical field; the universities and industry follow a close second.

Table 1.6 Advertised posts for microbiology graduates in 1973

Subject area	University	Research assistants	Student-ships	Post-doctoral fellowship	Hospital	Government	Industry
Plant pathology, mycology or virology (64)*	10	9	14	6	2	15	8
Animal virology (95)	13	14	13	12	10	19	14
Microbiology including medical virology (349)	72	39	35	40	30	70	63

*Total number in parentheses.

Table 1.7 Comparative distribution of advertised posts between 1961 and 1973

Subject area	1961	1969	1973
Plant pathology/mycology/virology	84	110	38
Animal virology	6	39	56
Medical microbiology including medical virology	140	233	239

Comparing the distribution of advertised posts in all sectors for 1961, 1969 and 1973 (*Table 1.7*) Dent and Caldwell revealed a rise and fall in the number of jobs for plant virologists, a continued increase in demand for animal virologists and a rise, then a levelling off, of jobs in medical microbiology.

Our own survey, which was more recent but not so exhaustive as the last, examined job opportunities for virologists advertised in *Nature (London)* and *New Scientist* over the latter six months of 1981. We did not include posts for molecular biologists, biochemists or immunologists unless viruses were a major part of the advertised research project. Surprisingly, in nearly all cases, jobs for microbiologists were essentially for bacteriologists. Whilst the survey showed an obvious bias for jobs in the UK (*Nature* and *New Scientist* being UK-based journals), the status of these journals, particularly the former, attracts employers internationally.

In all, within the six-month time span, 75 posts in virology laboratories were advertised, with the bulk (49) being in the UK, some in the USA (13) and the rest in Australia, Canada, Germany (FR), France and New Zealand. Out of the total, 64 were in the field of animal (including human) virology, 11 in plant virology and only 1 in bacterial virology. (This confirms the demise of the bacteriophage as a research tool.) The largest employers were the universities, accounting for 36 of the vacancies, many of which were fixed-term, usually three-year, research posts. The next largest employers were the governments, including hospitals (15), followed closely by the institutes and commercial sector (12 each), with many of the latter jobs in the developing applied fields such as recombinant DNA and hybridoma technology. The top salary in this survey was about £15 000 as head of a department in a commercial company (although not all the advertisements specified salary) and the lowest was just over £3000 for a junior technician. The average, however, was between £7000 and £8000.

Finally, of most interest, were the minimum qualifications required of the advertised posts. Nearly half (36) of these demanded a PhD, usually with some postdoctoral experience; over a third (28) required graduate qualifications (a good honours often specified); and only a small fraction (11) required HNC (Higher National Certificate, a British technical qualification) or less. It should be borne in mind that many of the very junior posts in virology laboratories would be advertised in the more local literature, such as newspapers.

What clearly emerges is that virology is very much the domain of the university-based short-tenured postgraduate on a none too attractive salary.

References

Anon. (1977). Employment news. *The Biologist* **24** (1) : 28. *Bergey's Manual of determinative bacteriology*. 8th edn. R. E. Buchanan and N. E. Gibbons, co-editors. Baltimore: Williams & Wilkins.
Copp, D. J. B. (1981). The state of biology in universities and polytechnics. *The Biologist* **28** (3) : 139–45.
Dent, A. J. and **Caldwell, I. Y.** (1974). Posts for biologists. *The Biologist* **21** (2) : 78–80.

1.5 THE INFORMATION NEEDS OF VIROLOGISTS

Though much is known about the information needs of scientists in general (mainly gleaned from citation practices), little is known about the needs of virologists or indeed microbiologists. To fill this gap in our knowledge it was necessary to mount a survey, and the results of this survey are the basis for the discussion that follows. Because of the relatively small size of the sample (20) and its UK bias the results should be regarded as being indicative, rather than truly representative.

Most virologists are agreed that their main sources of information are journals and personal, informal discussion; they are split almost equally as to which of the two is more important. Conferences, staff meetings, books and preprints are on the whole not major sources of information in terms of volume. There are, however, some important variations between kinds of virologist, with lecturers thinking more highly of conferences as information sources, researchers more highly of preprints and practitioners more highly of books.

With exception virologists think that to perform their job effectively it is absolutely essential *regularly* to seek out printed information (i.e. books, journals, reports). The frequency with which information is sought varies between weekly and monthly; few scan the literature less frequently than this, and few any more frequently. An exception is the more popular general scientific press, which is read more regularly on the whole. The major motivation for examining the literature is to keep abreast of general developments in the field. What triggers off the information search varies and includes, intriguingly, the arrival of the salary slip in the case of one respondent.

Whilst virologists obtain much of their information informally or orally (i.e. at conferences, staff meetings or in conversations over the phone or lunch), few see it as a substitute for the published literature, its unreliability being cited as the reason.

Within the virology community there is a general consensus that it is, today, virtually impossible to be fully aware of all the information that might be relevant to one's activities, although active researchers feel somewhat more confident in this respect.

Perhaps surprisingly, in a field that is thought to be truly international, the majority of virologists do not see it as being very important to be informed about similar work going on in the rest of the world — 'helpful but not crucial' is the widely held opinion.

There is a generally held feeling amongst virologists that there is too much being written on the subject of virology. Whether this is because a great deal of trivia are being produced, as some suggest, or because workers cannot cope with the sheer volume being produced, cannot be easily determined. It is often the case that overproduction occurs in some areas with possible underproduction in others. Like most sciences virology suffers from trends (at present towards molecular biology), so the subject of overproduction is likely to change with time.

Time and accessibility are obvious constraints on virologists' reading time, for all felt they could usefully use more information.

Whilst virologists do as a group generally like reading the literature, few enjoy searching through libraries (although they all appear to do it — at least once a week) and none seems to like examining bibliographies and abstracting and indexing journals.

Virologists would appear to be avid readers of journals; no doubt the very scatter of information in the field forces them to be so. The average number of journals scanned, at 9, is high by the standards of science; 15 is by no means uncommon.

The interdisciplinarity of the field is again highlighted by the fact that the average virologist belongs to three societies, with the particularly keen, who tend to be those in administrative posts, belonging to as many as seven or eight.

2 Organizations and their role in virology

2.1	**Introduction**	22
	Categories of organization	22
	Organizations as information sources	23
	Organizations as publishers of information	23
2.2	**Finding out about organizations**	23
2.3	**Organizations in virology**	26
2.4	**International organizations**	28
	World Health Organization	28
	International Union of Microbiological Societies including International Committee on Taxonomy of Viruses	33
	International Association of Biological Standardization	34
	Federation of European Microbiology Societies	35
	European Association against Virus Diseases	35
	International Organization of Citrus Virologists	35
	Commonwealth Agricultural Bureaux	36
	Commission of the European Communities	36
	International Agency for Research on Cancer	37
	International Comparative Virology Organization	37
2.5	**Commercial companies**	39
2.6	**National organizations — UK**	41
	Government bodies (Agricultural Research Council; Ministry of Agriculture, Fisheries and Food; Medical Research Council; Public Health Laboratory Service; Natural Environment Research Council; National Institute for Biological Standards and Control)	41
	Professional associations (Society for General Microbiology)	46
	Universities and colleges	47
	Other organizations — cancer research bodies (Cancer Research Campaign; Imperial Cancer Research Fund; Institute of Cancer Research)	50

2.7 National organizations—USA	51
Government bodies (US Department of Health and Human Services; National Institutes of Health; US Department of Agriculture; Department of Defense)	51
Professional associations (National Academy of Sciences; American Society for Microbiology)	53
Universities and colleges	54
2.8 National organizations—other countries	58
Australia	58
Canada	59
France	60
Federal Republic of Germany	60
India	60
Japan	61
USSR	61

2.1 INTRODUCTION

This chapter provides a comprehensive review of the various professional, academic, research and government agencies that together, admittedly sometimes unwittingly, provide for the well-being and development of virology. (Libraries are listed separately as directory entries 552–95, starting on p. 217.) The commercial firms that operate within the field are also dealt with, but the treatment is necessarily less comprehensive as there are many more of them and only a few make any meaningful contribution to the subject at large. As an indication of the number of organizations active in the field, over 600 are included in the WHO's *World list of virus laboratories*.

A myriad of interlinking organizations service the field: some provide funds for research, others conduct the research themselves; some provide reference facilities, whilst others actively disseminate information; some provide administrative support to run conferences, short courses, etc., others are policy-making or standard-setting bodies; finally there are those that are concerned with education and training. However—and this is the importance of a thorough knowledge of the organizations that so obviously abound in this field—most provide a unique and particular blend of services, and all may be usefully tapped for information.

Categories of Organization

Organizations are traditionally classified into four groups: professional associations, academic institutions, government agencies and commercial firms, with research institutes being something of a hybrid, sitting rather uncomfortably astride these four. Whilst such a classification can be helpful in bringing together organizations of common purpose or shared goals and, additionally, has the undoubted merit of being easily understood, it can be equally helpful—perhaps more helpful from an information-seeking standpoint—to see organizations from the viewpoint of the facilities they offer.

To illustrate: as far as information seekers are concerned (in theory anyway), it is a piece of information or a particular service they require; the *type* of organization providing it is at best marginally important and often is completely immaterial. Unfortunately however, in practice many people's information horizons are determined by the particular environment in which they work, be that academic, professional (in the case of practitioners) or government.

Organizations as Information Sources

It is plain that in the same way that a journal or book is a source of information so too is an organization, or perhaps more accurately, the people who represent the organization. Indeed an approach to an organization, whether by letter, phone or personal visit, is often likely to yield information of a higher quality more quickly and easily than the conventional literature search. Furthermore, the information obtained is likely to be more current and the act of obtaining it far less onerous. Not surprisingly this particular method of obtaining information is exceedingly popular, particularly with practitioners whose time is at a premium — many of whom have built up, over a period of time, a network of contacts to whom they may refer for particular kinds of information.

Many organizations also act as information clearing houses directing the user either to alternative or to additional sources of information — a particularly useful facility when one considers that, in the initial stages of the information search anyway, most people's conceptions of what they need are necessarily vague (it is not easy to decide what is needed until one is aware of what is available).

Organizations as Publishers of Information

Most organizations are, of course, also prolific publishers of information and it is a fact that in virology a good deal of information emanates from outside the traditional commercial publishing channels. Thus if we take as an example journal publishing we have the following distribution: approximately 40% of journals are published by societies, 5% by academic institutions, 2% by government and 53% by commercial publishers. The fact that this latter group accounts for over half of all journals published suggests that publishing in the field has its financial attractions.

We recognize intuitively that the characteristics of the publications issued by the aforementioned groups might well differ by referring to certain documents as being 'academic' or 'commercial'. Thus the expression 'academic' not only sugests that it is a publication of a college or university but it also conveys much about the intellectual level, format and purpose of the journal. In the case of the ubiquitous scientific journal the differences, although visible, might not be all that marked but in the case of books and reports the differences are quite evident.

2.2 FINDING OUT ABOUT ORGANIZATIONS (*see* bibliography entries 1–15)

When it comes to locating and finding out about virology organizations, then the

World list of virus laboratories (5th edn, 1979) is the essential reference source. It is basically a list of laboratories active in the field, but because it includes details of the organizations to which these laboratories are attached and because these organizations include universities, public health authorities, hospitals, research institutes and government agencies its use goes well beyond its limited aims. It is of course really a guide to those organizations that conduct research into viruses.

Its stated aim is to promote the exchange of information within the framework of technical cooperation in the field of virology on national, regional and international levels. The *List*, which is published about every five years, provides in tabulated form a record of the activities of 616 laboratories located in 89 countries. Prepared by the Virus Diseases Unit of the WHO, the record is naturally an authoritative, accurate and comprehensive one. In addition to the address the following details are provided for each laboratory: head; type of activity (i.e. hospital, research, teaching, public health); number of professional staff; and field of activity (i.e. which viruses — 11 groups are named — are investigated). In addition the record denotes whether the laboratory cooperates in the WHO Virus Disease Programme as a National Influenza Centre or Virus Collaborating Centre for Reference and Research.

The *List* is arranged according to WHO Region in alphabetical order. Within each Region the order is by country, again arranged alphabetically. In general, the laboratory with the functions of a central public health laboratory or national laboratory is listed first and is then followed by other laboratories arranged alphabetically by towns or cities. The exception is the entries for the USA and Canada, which are sub-grouped according to province or state.

To judge from this *List* the US would seem pre-eminent in the field with 97 organizations involved. Next, by some distance, comes the UK with 82 and then the Federal Republic of Germany with 41.

If it is a question of tracing colleges and universities that teach virology then we need to look to three publications, one American and the other two British. Undoubtedly the US publication, *Directory of colleges and universities granting degrees in microbiology*, is the most useful and impressive of the three. Published in 1980 by the American Society for Microbiology, the *Directory* is divided into three: section 1 lists departments granting degrees in microbiology; section 2 lists individual schools and contains current information such as courses being taught, special approaches to teaching and areas of concentration; section 3 is the most useful as it enables the reader to identify schools that offer specialist courses (i.e. virology).

To obtain similar, although not as detailed, information for the UK one needs to look at two publications: the annual UCCA-published *How to apply for admission to a university* and the somewhat more irregular *Graduate studies*. The first is a guide to universities showing which first-degree subjects they offer; a subject index enables one to locate those universities offering virology degrees. Thus the Manchester BSc in Bacteriology and Virology and the Warwick BSc in Microbiology and Virology are mentioned. Those universities offering a higher degree in virology can be located in *Graduate studies*, where they are listed under the heading 'Microbiology'. A fairly detailed explanation of the content of each course is offered.

Identifying organizations that research in the field of virology is relatively straightforward in the case of government and academic research but not so easy for commercial companies, where problems of confidentiality arise. In the latter case it is probably best to consult the company's annual report or research in progress bulletin, if they have one.

For keeping in touch with virology research in the US and Canada, *Research center directory* and its updating service *New research centers* are indispensable tools. The *Directory* is published every three years and is essentially a guide to research in universities and independent, non-profit-making organizations. These bodies are arranged in broad subject categories with the one headed 'Life sciences' being of most interest. It is possible to identify virology research either by browsing through this section, or far better, by using the subject index. The importance of using the index is brought home by the fact that whilst only 10 virology projects are listed in the life sciences category another 34 are scattered throughout the other subject sections — demonstrating more than anything else the interdisciplinary nature of the subject. Unfortunately many of these are of peripheral interest but as there is no way of telling this from the index entry 'Virology' it is necessary to scan them all.

Each project is described in about 12–15 lines giving amongst other things names of those involved, addresses of the institution, a brief guide to content and publications.

The basic directory is updated by the quarterly *New research centers*. Whilst this provides no subject index the number of entries is probably small enough to allow browsing, certainly if the search is restricted to the 'Life sciences' section; although as pointed out above this might be unwise given the scatter of projects throughout the classification.

A less conventional approach to listing US research in progress is offered by the Smithsonian Science Information Exchange (SSIE) computer database, which can be accessed via DIALOG. The database contains summaries of research either in progress or completed in the last two years. Project descriptions are received from over 1300 organizations, mainly government agencies, but academic and private organizations are also contributors. The database contains about 180 000 citations to research projects and is updated monthly. SSIE covers the biological and medical sciences in some depth and there were 7900 citations to virology projects as of March 1982. The abstracts that accompany each citation are particularly valuable, frequently over 200 words in length.

Within the UK, fair coverage of research in progress is provided by two publications: one, *Research in British universities polytechnics and colleges*, provides a listing of academic research and the other, *Industrial research in the United Kingdom*, provides a comprehensive picture of research in industrial firms, public corporations, research organizations, government departments and academic bodies.

The coverage of *Research in British universities polytechnics and colleges*, an annual directory published by the British Library, is self-explanatory. Volume 2, *Biological sciences*, is of prime interest. Research projects are listed under their respective host institutions and these are in turn grouped according to their subject concerns. Whilst many virology projects are listed under the heading 'Microbiology' a sizeable number are scattered throughout the volume, chiefly under 'Biochemistry', 'Biology' and 'Veterinary science'. To locate all these one needs to use the subject index.

Industrial research in the United Kingdom is a frequently updated directory (latest edition 1980) of all the organizations involved in research, not just the academic ones (it also provides details of learned societies and trade associations).

Organizations are grouped by their types (e.g. government, industrial firms). Each entry is accompanied by a detailed profile indicating the size of the organization, its main research and development personnel, major research interests, annual R & D budget and specialized research facilities, though entries are not always as informative as this, those for commercial firms being somewhat sparse. There is a subject index and the entry there for virology leads us to four government agencies. However, many of the other organizations do have virological interests but these interests are not always specified, being incorporated within a statement about a more general field of interest.

Tracing organizations providing grants for research and study in virology is made relatively straightforward by a database called GRANTS; this is available on DIALOG. GRANTS, produced by Oryx Press, provides information on 2200 available grants offered by government, commerce, private foundations and professional associations. Organizations are listed as offering grants in the general area of virology. Included in each entry is information about money availability and qualifications required. The GRANTS database corresponds to the printed publications of the Grants Information System (GIS), the *Quarterly Cumulative* volumes and the *Faculty Alert Bulletin*.

Another useful publication in this connection is the *Grants register*, which provides details of research grants, vacation awards and travel grants available throughout the USA, UK and Commonwealth. A subject index enables virology grants to be traced (the virology entry is under the 'Medical and health sciences' heading). Four organizations — the John Innes Foundation, the Imperial Cancer Research Fund, the Institut Pasteur (Paris) and the Overseas Development Administration — are listed as providing funds specifically in the field of virology. Each entry includes the following details: purpose, value, duration, eligibility and application dates of the award. Appearing every two years (the latest edition is the 7th, 1981-3), the *Grants register* is invariably less current than its American counterpart, GRANTS.

The few societies and professional organizations that exist in the field can be traced in the following reference works: *Encyclopedia of associations* (American), *Trade associations and professional bodies of the UK*, *Directory of British associations*, *Directory of European scientific organizations*, *World guide to scientific associations* and *World of learning*. They all have subject indexes which enable virology and microbiology societies to be identified. The *Encyclopedia of associations* is the largest and most detailed of them all, covering 15 000 organizations and providing an abstract of the scope, purpose and major activities of each organizations; it is available in hard copy and machine-readable form (via DIALOG).

For tracing international organizations in virology the *Yearbook of international organizations* is particularly useful (a detailed subject index providing access to strictly virology organizations) and authoritative (it is published by the Association of International Organizations).

2.3 ORGANIZATIONS IN VIROLOGY

An initial glance at a list of organizations that have assumed responsibility for variously organizing, administering, financing and publicizing virology will give the impression that the field is complex and diffuse, the responsibility being dispersed amongst many and varied hands. Closer examination will confirm the complexity but the trained eye will detect that a high degree of interrelatedness and cooperation exists between the various, apparently unconnected, bodies. The connections between these bodies are most obviously strengthened by the fact that they have members in common. It is not uncommon, for instance, for an eminent virologist to be a member of an educational faculty, a research association, a national microbiological society and a medical or veterinary association; adviser to a government scientific review committee; possibly on a WHO consultative committee; and almost certainly on the editorial board of a couple of virology journals. Cooperation between organizations is further promoted by the fact that national, commercial or even ideological boundaries appear to have little constraint on the free exchange of information between virologists, this no doubt being due to the fact that virology has proved, to date anyway, to be remarkably free from the military and commercial rivalries that beset many subjects.

One further point illustrates the intricacy of the relationship between virological institutions: commonly the organizing of conferences or symposia is a cooperative effort. Thus a national microbiological society will organize a meeting, with finance coming from the responsible government agency or an international body such as the WHO, and facilities being provided by a university department. From an information-seeking point of view such practices can lead to considerable confusion when it comes to establishing precisely who is responsible for the publication of the proceedings.

As yet there are no major independent societies that are devoted to virology; the effective group remains very much at the microbiological level. (Japan is the exception, where there is the Society of Japanese Virologists.) Virologists thus share the forum with bacteriologists and, usually, mycologists. It is probably true to say that bacteriology is the dominant concern of most microbiological societies. However, the number of meetings where viruses are the sole subject under discussion has expanded enormously over the past decade, albeit that they are sometimes disguised under such titles as chemotherapy, vaccination, interferon and immunology. The implication of this latter fact is that much communication in the subject is conducted outside the immediate field of view of the virologist (that is, it is not visibly labelled as virology).

The workplaces of virologists, like those of workers in many other scientific disciplines, are universities, government or quasi-government research bodies, privately run non-profit-making research institutes and the research departments of commercial firms, commonly drug companies. The relative strength of these sectors varies considerably from country to country. This point can be nicely illustrated by analysing the contents of virological journals to ascertain the occupations and workplaces of the contributing authors. Six journals were surveyed over the period 1980–1: *Virology*, the *Journal of General Virology*, the *Journal of Virology*, the *Journal of Medical Virology*, *Intervirology* and *Archives of Virology*. The result of this exercise is summarized in *Table 2.1*. It can be seen from this table that in France virus research is largely conducted in government or quasi-government agencies, whereas in Canada and Japan research is predominantly the domain of the universities. This is also partly true of the USA and the Federal Republic of Germany, although in the latter country a significantly large proportion (over one-third) of research is carried out in autonomous research institutes. Most countries have a very active university contribution, apart from

Table 2.1 Workplace of authors of articles in virology journals

Type of organization worked for	Country of origin					
	USA	UK	Japan	France	Germany (F.R.)	Canada
Government, including research councils	25.5%	48.0%	18.0%	70.0%	–	15.0%
Private non-profit-making institute	10.0%	16.0%	9.0%	17.0%	39.0%	9.0%
Commercial	1.5%	4.0%	5.0%	–	–	–
University and medical colleges	63.0%	32.0%	68.0%	13.0%	61.0%	76.0%

France, where participation is surprisingly small. In all countries the commercial sector would appear to play an insignificant role in virus research. It should be pointed out however that the particular method of analysis used to determine organizational representation probably underestimates the role of virologists working in the commercial sector. After all, academics have a built-in advantage when it comes to publication as they are virtually paid, certainly given the time, to write. (It is also likely that it is not always in the best commercial interest to publish one's work.) Whatever the reason, commercial virologists' role will inevitably grow as work expands on the applied/technological aspects of the field such as biotechnology and monoclonal antibody production.

2.4 INTERNATIONAL ORGANIZATIONS (see directory entries 393–404)

World Health Organization (WHO)

Function and organization

The role of the WHO, a United Nations specialized agency, is a central one in virology. It has had a long-standing interest in the subject, as witnessed by the fact that it was represented at the very first Congress for Virology back in 1968, when it delivered a paper on its particular role and responsibilities in the field. Furthermore, it is undeniable that the greatest advances in our knowledge of viruses have been made since the Second World War and it may or may not be a coincidence that this period also encompassed the birth of the WHO (1947) and the rapid development of its activities.

The WHO's prime function is the very practical and worthy one of improving the health of mankind as quickly as possible; virus diseases have of course constituted a major obstacle to this objective. Inevitably then it is the medical aspects of virology that most benefit from the efforts and financial support of WHO.

Not unnaturally the WHO is principally concerned with the problems that individual countries cannot solve with their own resources. Smallpox is a case in point: the eradication of this disease marks a highly significant phase in virology's development and can be largely attributed to the international cooperation that is embodied in the WHO. Today the public health and medical professions of more than 150 countries exchange their knowledge under its auspices.

The WHO's work is largely conducted through a number of advisory groups and committees. *Figure 2.1* provides a graphic representation of the advisory group/committee structure; those with a particular viral interest have been made prominent. To each division within an advisory group a number of Expert Advisory Panels report; the ones concerning virology are those for Biological Standardization, Cancer, Rabies, Immunology, International Surveillance of Common Diseases, Zoonoses and Virus Diseases. Virus diseases are also the concern of a WHO Scientific Group.

Activities of the WHO that concern virology

Referral services. A network of international and regional reference centres, based upon existing national laboratories, has been fashioned by the WHO to provide a comprehensive reference service covering the areas in which the WHO is directly involved; this of course includes virus work. In addition members of the network maintain a virus bank and run training courses.

The WHO's *World list of virus laboratories* (1979) (bibliography entry 13) includes a total of 51 such centres. Just under half are described as Virus Collaborating Centres or Centres for Virus Reference and Research. The rest specialize in research and reference work in connection with arboviruses, smallpox, influenza, hepatitis, respiratory and enteroviruses, simian viruses, rotaviruses, special pathogens such as Marburg virus, food virology, rapid laboratory virus diagnosis, and comparative virology. The names and addresses of these laboratories are listed as directory entries 479–529.

Research. The WHO is involved in research in two ways: directly, in the sense that it has its own laboratories, and in a rather more indirect way by funding research through a large number of collaborating laboratories. In all, well over 500 laboratories are engaged in WHO-sponsored work. *Table 2.2* shows how these laboratories are geographically distributed.

As the WHO would be first to admit, it is ironic that most of the research takes place in the better-developed parts of the world where the problems of disease are not so severe. It is true of course that much of the research done in the developed countries concerns diseases in the Third World.

Reagents programme. The reagents programme concerns the provision of standardized preparations for research and diagnostic laboratories. Prototype virus strains and reference antisera of nearly all the major diseases such as measles, enteroviruses, hepatitis, rubella and herpes viruses are available from WHO-associated laboratories. In addition, cell cultures needed for virus isolation may be supplied, but generally speaking, laboratories are encouraged and assisted

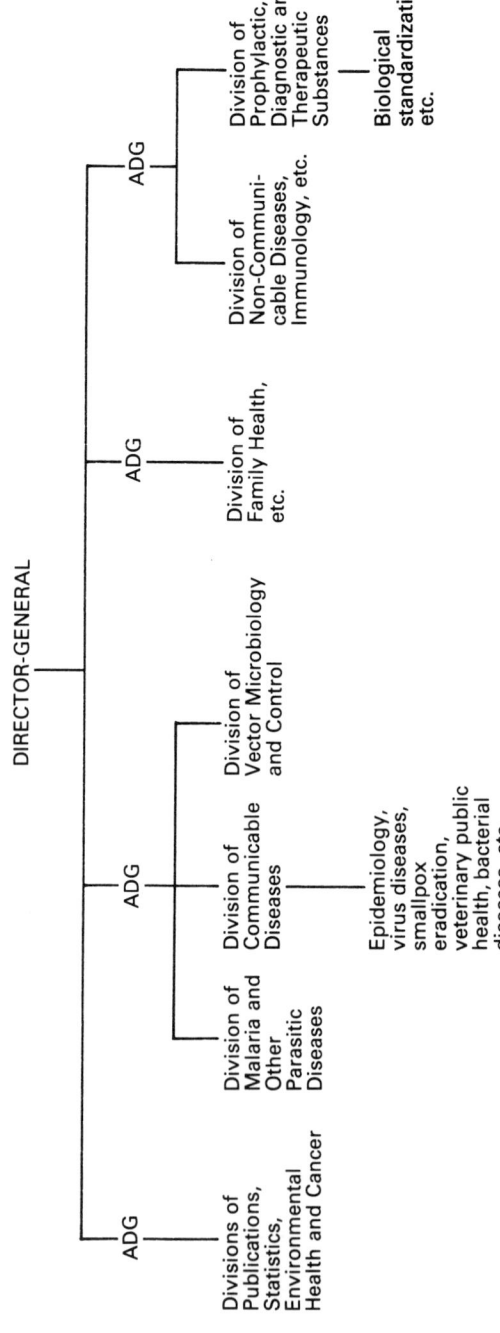

Figure 2.1 Structure of the organization of the World Health Organization's headquarters secretariat

Table 2.2 Geographical distribution of WHO laboratories

WHO region	Number of countries represented	Number of laboratories
Africa	11	17
America	19	149
Asia, south-east	6	24
Europe	29	285
Mediterranean, east	10	18
Pacific, west	11	45

to develop their own strains. The WHO Expert Committee on Biological Standardization is served by an international group of experts who publish reports regularly through the WHO's *Technical Report Series*.

WHO team for special studies in Africa. To redress the balance in favour of the developing countries, a programme centred on the East African Virus Research Institute at Entebbe was initiated to study virus diseases such as polio and those diseases caused by the respiratory syncitical virus and adenoviruses. Political problems in Uganda have unfortunately disrupted its programme.

Collection and dissemination of documentary information. The WHO virus reporting system, established as long ago as 1967, has amassed a data bank of well over a quarter of a million reports on viral infections. One hundred and nineteen laboratories participate in the scheme. The objectives of the virus information system, according to Assaad and Bres (1977), are

1. 'Surveillance of viral diseases: to obtain regular, systematic information on the distribution of particular virus infections in different parts of the world; to promptly recognize changes in their distribution; to obtain broad indications of the main clinical manifestations and age groups affected.

2. 'Assistance in disease control: to alert local, national and regional health authorities to changes in disease distribution (particularly epidemics) that have more than local significance and require coordinated investigation and containment operations; to provide these authorities with data on the distribution of virus infections in their own and neighbouring areas that may assist them in planning control programmes; to coordinate vaccine surveillance (safety and efficacy).

3. 'Study of patterns of virus behaviour: to recognize differences in the pattern of virus infection in communities with various social, economic, and geographical conditions and to follow the effects of changing environmental conditions which may assist understanding of their ecological behaviour; to further

investigate data obtained from reports of routine diagnostic virology by promoting planned studies based on defined populations (e.g. patterns of immunity and frequency of inapparent infection).

4. 'Exchange information: to offer a forum for exchange of information on local experiences in the investigation and control of virus infections and the use of new laboratory or other investigative techniques.

5. 'Data collection: to encourage the systematic collection and reporting of accurate statistics locally and at a national level; to show the value of consolidated reports to local laboratories, physicians, and health administrators in interpreting isolated cases and in understanding the behaviour and directing the control of diseases in their areas.

6. 'Laboratory methods: to promote the development and trial of new techniques; to encourage the critical evaluation of laboratory performance.'

The data flowing into the system are processed to produce quarterly and annual reports, annual reviews of specialist areas (e.g. enteroviruses) and ad-hoc retrieval requested by laboratories.

'Ad-hoc retrieval of information may be requested by any participating laboratory. It may be restricted to information provided by the laboratory or may extend to the entire information available in the system. An example of the former was that of one laboratory requesting detailed analysis of enteroviral infections by age. Another example is that of one laboratory asking for the list of laboratories which undertook typing of herpes simplex virus at a time when herpes was not frequently reported by type. The purpose of the inquiry was to contact other laboratories with a similar interest to exchange detailed information.' (Assaad and Bres, 1977).

The WHO in collaboration with the FAO was also instrumental in the setting up of a body — the International Comparative Virology Organization — whose specific function it was to collect and evaluate virological data. The activities of this organization are dealt with separately (*see* section 2.4).

Publishing. The WHO is also a prolific publisher. The major vehicles for disseminating information on its current work are: the *Bulletin of the World Health Organization*, a scholarly journal that regularly reports on the current status of viral diseases worldwide (e.g. 'Morphology and morphogenesis of arena viruses', **52** (1975) 409–19); the *WHO Chronicle*, a monthly newsletter containing general information on the principal activities of the WHO, including news on conferences, meetings and funding; and *The Work of the WHO*, which is the official biennial record of its activities.

Much of the WHO's past research and missives on standards and techniques are embodied in two report series: the *Monograph Series*, of which 2–3 per year may be published; and the *Technical Report Series*, of which a somewhat daunting 200 per year might be issued. The *Technical Report Series* also includes the official reports of the work of the various Expert Committees. Examples of the kinds of title that are published in these report series are: *Laboratory techniques in rabies* (3rd volume of the *Monograph Series*, published in 1975) and 'The use of viruses for the control of insect pests and disease vectors' (*Technical Report Series*, published 1975).

The indispensable *World Health Statistics*, a monthly publication, cumulated annually, contains statistics on notifiable diseases as well as general data on infant mortality, etc. Such a service could come only from the WHO because its production is an international undertaking that requires considerable national cooperation.

A complete record of the WHO's publishing activities can be found in the *World Health Organization publications catalogue, 1947–1973* and its five-year supplements.

International Association (Union) of Microbiological Societies (IUMS)

Founded in 1930, with its head office in Marseilles, the IUMS embraces 49 national microbiological societies, of which the Society of General Microbiology and the American Society for Microbiology are particularly active members. The IUMS, then, is largely a forum for the international exchange of ideas and information, and much of its work is administrative and regulatory in nature.

Until quite recently the IUMS was a subdivision of the International Union of Biological Sciences. However, in 1978, at its Munich Conference, the IUMS decided on a parting of the ways; mainly because this would, it was thought, boost its standing in the UK, where it would now perhaps be entitled to national (i.e. full) rather than subcommittee status within the Royal Society. In 1980 the then International Association of Microbiological Societies declared itself a union in its own right. The decision as to whether the scientific community will recognize its new-found status is still pending.

The IUMS's specific responsibilities lie in organizing international microbiological congresses and supervising the work of 40 specialist committees and commissions, of which the International Committee on Taxonomy of Viruses (ICTV) and the Commission for Biological Standardization are of particular viral significance. The way in which the IUMS is structured is represented graphically in *Figure 2.2*. The three bodies in bold type have special significance for virology and their activities are dealt with in detail below.

Figure 2.2 Organizational structure of the International Union of Microbiological Societies

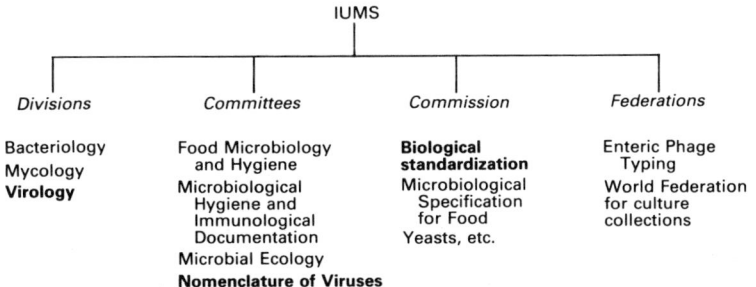

Virology Division of the IUMS

The IUMS has fully recognized the legitimacy of virology's claim for independent subject status by creating a separate section, devoted entirely to the pursuit of matters virological. The decision to form a separate section was in fact taken as far back as 1966, at the IUMS's (then IAMS) 10th International Congress for Microbiology in Moscow. The official mouthpiece for the Virology Division is the journal *Intervirology*.

International Committee on Taxonomy of Viruses

One other subordinate body of the IUMS deserves special mention, the International Committee on Taxonomy of Viruses. As its name suggests, the Committee is concerned with taxonomy and the standardization of nomenclature in virology. The work of the Committee is in fact conducted by seven subcommittees: there are specialist committees for bacterial, fungal, invertebrate, plant and vertebrate viruses, the two remaining committees being largely administrative in nature. The membership of the committees is drawn from the national societies affiliated to the IUMS.

Perhaps the most visible products of the Committee's deliberations are the fairly regular updates on viral taxonomy. They are usually published in the journal *Intervirology*. The most recent appeared in 1982 and constituted the fourth report.

International Association of Biological Standardization

The International Association of Biological Standardization (IABS), a commission of the IUMS (albeit a very independent one), is primarily concerned with the development and use of standardized techniques in the biological sciences. In this it has a common concern with the WHO — standardization is after all an international issue — and as a consequence there is a good deal of collaboration between these two international organizations.

The importance of the IABS can be best judged by reference to the fact that nearly every meeting it has held since its very first in 1955 has had some relevance to virology, admittedly a virology seen largely from a viral vaccination and chemotherapy point of view. Subjects treated have included rabies, foot-and-mouth, rubella, smallpox and influenza. The most recent congress (the 17th, held at Lyon, 1981) was entirely devoted to herpesviruses.

The IABS organizes three types of scientific meeting: first, International Congresses, which take place every two years and cover a wide range of topics; secondly, Symposia, which are held two to three times a year and are concerned with single, much more specific, issues; and thirdly, the annual Meetings of the Committee on Human Diploid Cell Strains, at which the use of cell lines for vaccine production is reviewed.

In addition to organizing conferences and publishing their proceedings the Association publishes a quarterly journal called the *Journal of Biological Standardization* and a bilingual newsletter, *Bulletin d'Information/Newsletter*, which is also published quarterly.

It is highly probable that since 'standardization' is a neglected and not highly thought of field of virological study (perhaps because it is largely applied), the work of the IABS escapes the attention of many virologists.

Federation of European Microbiology Societies

In 1979 the Federation of European Microbiology Societies (FEMS) replaced the North-West European Microbiological Group as the regional organizer for microbiology in Europe. However, despite its grand title and seemingly important function it is not a particularly active body, most of the work for which it would seem responsible being undertaken by the national societies and the IUMS.

The geographic field of interest is interpreted fairly liberally, with such countries as Turkey, Israel and the USSR being counted amongst its 21 members. By far the largest national contributor to FEMS' budget is the Society for General Microbiology (SGM), which in fact was largely responsible for its establishment. The relationship goes even further than this, for the SGM's two journals — the *Journal of General Microbiology* and *Journal of General Virology* — are also the official organs of FEMS.

FEMS does however have a small publishing programme of its own, issuing two periodicals: one a newsletter simply called *FEMS*, which lists officers of the Federation, publications and forthcoming meetings and is usually available as an insert to the above-mentioned journals; and the other a journal called *Microbiology Letters*, which is essentially a vehicle for short reports on current microbiological research. Additionally FEMS organizes regular conferences and publishes their proceedings but as yet none has featured virological topics.

European Association against Virus Diseases

The European Association against Virus Diseases is one of the few organizations to deal exclusively with virology. Established in 1951 in Geneva, following conferences held in Brussels (1948), Paris (1949) and Amsterdam (1950), the Association has broadened its outlook over the years. The expanding interests of the Association are in fact marked by a number of name changes: thus it started life as the European Association against Poliomyelitis and Allied Diseases, changed to the European Association against Poliomyelitis and Other Viral Diseases and ended up with its present all-encompassing title.

Today, as its current title suggests, the Association is concerned with known virus diseases, largely from a clinical and epidemiological point of view. It pursues this interest by promoting international collaboration between bodies sharing similar aspirations. It draws its membership of 22 mainly from European countries.

Perhaps the most tangible part of the Association's work is the conferences it organizes and the proceedings it publishes (these being of limited availability). One of the most recent was the 18th, held at the University of Stirling in 1981. Amongst the main topics discussed were monoclonal antibodies, enteric viruses, interferon and other antiviral agents.

International Organization of Citrus Virologists

The economic importance of combating viral disease in citrus fruits led to the founding in 1957 of this specialist, California-based, organization. Its stated aims are to promote research into citrus diseases and to place this research on an international footing. Members come from 38 citrus-growing countries, the UK of course not being represented. In addition to organizing conferences and publishing their proceedings the Organization also publishes a journal *Citrus Virus Diseases* and a number of handbooks, including *Indexing procedures for fifteen virus diseases of citrus trees* (US Department of Agriculture Handbook 333).

Commonwealth Agricultural Bureaux

The Commonwealth Agricultural Bureaux (CAB) are unusual in that in addition to conducting research they are a major information provider. Established as long ago as 1929 and sponsored by the various Commonwealth governments, the CAB are charged mainly with the task of disseminating agricultural information. Whilst strictly speaking they are a multinational organization the CAB are however largely UK-run and -based.

The CAB comprise 4 institutes, 10 bureaux and a headquarters at Slough. Of these the Commonwealth Mycological Institute at Kew, the Bureau of Animal Health at Weybridge and the Institute of Biological Control in Trinidad are of special interest to virologists.

The Commonwealth Mycological Institute was founded a decade before the CAB with the objective of collecting and disseminating information on the fungal, bacterial and viral diseases of plants. Amongst its publications are an abstracting journal, *Review of Plant Pathology*, and the indispensable *Descriptions of plant viruses*, of which there have been over 200 published. The latter is sponsored jointly by the Association of Applied Biologists (for more details *see* p. 89).

The Institute of Biological Control has been particularly active in studying the potential use of viruses to control pests.

The Bureau of Animal Health provides via its two abstracting services — *Index Veterinarius* and *Veterinary Bulletin* — a worldwide information service in the veterinary sciences. Three other abstracting services are also of relevance, those covering Animal Health, Medical and Veterinary and Plant Pathology. Its annotated bibliographies cover some 3000 subjects, which include many important viruses.

The CAB also provide a service by which individual users can conduct on-line searches of their computerized database containing over one million abstracts, of which most refer to journal articles.

Commission of the European Communities

Under the Directorate for Coordination of Agricultural Research, the Commission of the European Communities sponsors and organizes research into various aspects of animal disease. It has made a most notable contribution in the field of both bovine and avian leukosis, including Marek's disease — all caused by tumour viruses. The Commission runs conferences and their proceedings are published by the Office for Official Publications of the European Communities in Luxembourg. One such conference was Studies on Viral Replication, held in Brussels in 1974.

The Commission, in collaboration with the CAB, publishes *Animal Disease Occurrence*, which is a biannual veterinary serial containing approximately 200 abstracts and tables on geographical occurrences of the various animal diseases. It also, usefully, provides a list of research projects that it finances called *AGREP — Permanent inventory of agricultural research projects in the European Community*; virus projects are found under the following headings: 'Plant diseases and disease control'; 'Animal diseases, veterinary medicine'; and 'medicine'. A directory of personnel is also appended.

International Agency for Research on Cancer

The International Agency for Research on Cancer (IARC) was established in 1965 as an independent organization within the WHO framework and obtains the greater part of its financial assistance from the National Cancer Institute and the National Institutes of Health, both of the USA.

The IARC was specifically charged with the responsibility for cancer research. Under a special virus cancer programme, potential carcinogens such as oncogenic viruses are being studied. The headquarters are in Lyon, France with research centres in Iran, Kenya and Singapore.

The Agency, in collaboration with about 70 national research institutes, is also involved in a programme to advance education and training in the field of cancer research.

As well as numerous publications on chemical carcinogens, a ajor contribution to the virology literature was *Oncogenesis and the herpesvirus* edited by de Thé *et al.* (1978) (bibliography entry 262), which was the published proceedings of a conference on this subject. There have been two other such conferences.

International Comparative Virology Organization

In cooperation with the FAO, the WHO began to develop an international programme on comparative virology in 1967 which was to lead eventually to the founding of the International Comparative Virology Organization (ICVO) (*Figure 2.3*). The major function of the programme was to bring together the various strands of virological investigation with the desired objective of concentrating resources and effort on disease problems of man, domestic animals and plants.

The first international conference was held in Quebec, its proceedings being published as *Comparative virology* edited by Maramorosch and Kurstak (1971) (*see* p. 68). The momentum was maintained with a symposium sponsored by the Society for General Microbiology in the UK in 1972, another on Comparative Viral Immunodiagnosis in Montreal a year later and one in Viral Zoonotic and Epizootic Diseases of Veterinary Interest held at the Institut für Medizinische Mikrobiologie, Infektions- und Seuchenmedizin in Munich in 1975.

The Munich symposium led in the same year to the establishment of the WHO Collaborating Centre for Collection and Evaluation of Data on Comparative Virology, with offices and a laboratory near the Institute. The Centre is financed by the Federal Ministry for Youth, Family and Health of Germany (FR), and is in the process of collecting and evaluating data on animal viruses in order to establish a catalogue. This information is supplied by invited specialists, stored in computers on magnetic tape and will be available to national public health and

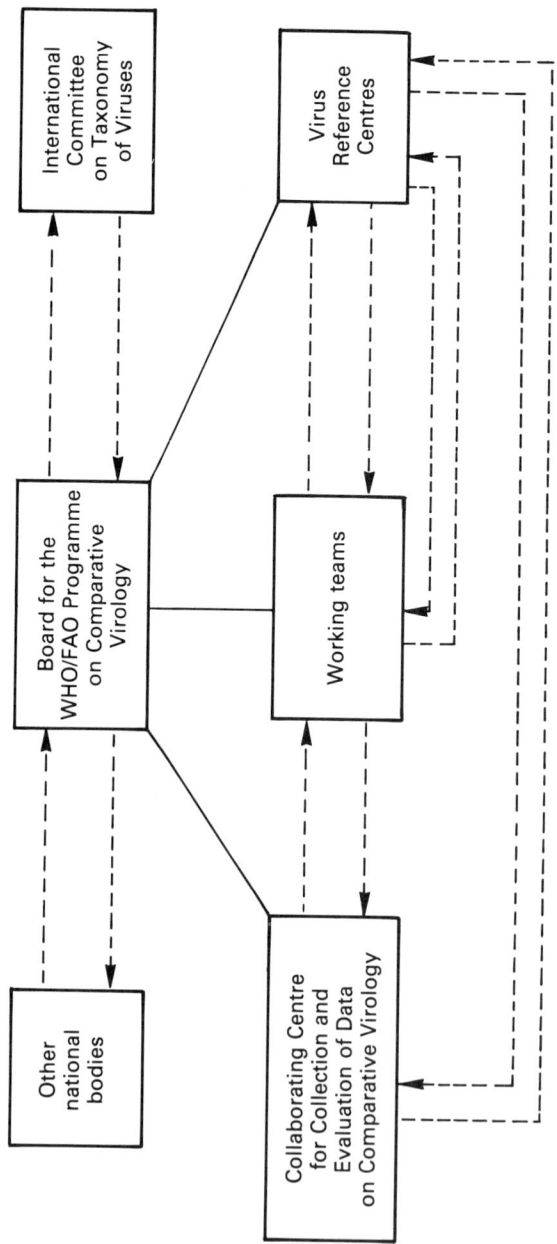

Figure 2.3 Structure of the WHO/FAO programme on comparative virology. (———) Line organization; (– – – –) functional relationships

veterinary authorities, institutions including libraries and scientists interested in comparative virology. The scheme, at the time of writing, is in the preparatory phase. A further function of the Centre has been the organization of the annual Munich Symposia on Microbiology. The 5th Symposium (1980) dealt with 'Comparative Aspects of Leukaemias, Lymphomas and Papillomas'. Its proceedings, edited by Bachmann (1980), were published commercially for the first time, thus ensuring a wider distribution (*see* bibliography entry 249).

In 1977, a decade after the first conference, the 4th International Congress finally established the ICVO. The stated functions are: to organize the International Conferences on Comparative Virology; support and develop the data collection; and create a *Review of Comparative Virology* and to publish books dealing with the subject.

The published proceedings of relatively recent symposia include *International symposium on comparative aspects of Arctic and tropical arboviruses* (1975) and *International conference on the impact of viral diseases on the development of African countries* (1976) (*see* bibliography entry 29).

Reference

Assaad, F. and **Bres, P.** (1977). The World Health Organization virus information system. In *Comparative diagnosis of viral diseases*; vol. 1, *Human and related viruses*, Part B. London: Academic Press.

2.5 COMMERCIAL COMPANIES

Because of the multinational nature of the major companies involved in virus research we shall treat them separately rather than subsume them to the countries that house their headquarters.

The contribution of the commercial companies to virology is difficult to gauge. In the applied areas of the science such as chemotherapy and vaccination the contribution is probably significant, yet because of a noticeable reticence in passing on the fruits of their research to the rest of the virology community, their contribution towards the literature is small. Few managements actively encourage the publication of results obtained by their staff even after patents have been obtained. Virtually all companies place heavy emphasis on gaining proprietary advantages through the work of their staff. Whilst this is to be expected to a certain extent, the cloak of secrecy that pervades this work is not in the best interests of virology, making it exceedingly difficult to find out what research is carried out. There are however companies that have contributed immensely to the advancement of virology either by allowing their staff, albeit a small fraction, to conduct publicly available research or by financing research in the universities and institutes; but, in general, sponsorship of basic research by industry leaves much to be desired (*see Table 2.4*, p. 49).

At the forefront of the commercial efforts in virology is the UK-based Wellcome Foundation, whose origins (as Burroughs Wellcome & Co.) roughly coincided with that of virology itself. Unlike many industrial companies it has not confined its activities solely to the commercial development of research findings; much of its research has been truly innovative. Expenditure on R & D for the year ending August 1981 totalled a massive £52 million, amounting to between 10 and

11% of total sales revenue. The principal activities of the Foundation, which comprises a total of 60 companies throughout the world, are the discovery, manufacture and sale of medical and veterinary pharmaceuticals, biological products, diagnostic reagents as well as other preparations. More specifically it maintains a leading position in both the production of interferon and the study of its clinical utility. Long-term work in antiviral chemotherapy has resulted in the introduction of a compound for use against herpetic infections. Wellcome has also been intimately involved with the work of the Animal Virus Research Institute at Pirbright — an association that in 1961 led to the building of a laboratory within the latter's perimeter to exploit commercially the live virus vaccine that was being developed by the Institute to combat foot-and-mouth disease. The Wellcome Pirbright Laboratory, now well established, publishes the *Foot and Mouth Disease Bulletin*, giving details of the world situation as regards this disease. Another publication, this time produced by Wellcome Reagents, a subsidiary, is *Lablore*, which is a small monthly review and abstracting service, mainly for those working in hospital microbiology and immunology laboratories.

Some of the profits from the Foundation are channelled into the Wellcome Trust, a registered charity, which in turn distributes money to aid medical and veterinary research in the universities and institutes; between 1978 and 1980 this financial assistance totalled £20 million.

The 150-year-old Smith Kline Corporation is a well-established multinational company, with headquarters in Philadelphia, which researches and develops (to the tune of $136 million a year, nearly 8% of total sales), manufactures and markets ethical pharmaceuticals, medicines and animal health products. The Corporation also provides medical laboratory services throughout the USA and Canada. In the UK, the Corporation has recently built a modern and very well equipped research institute, and created a foundation for the advancement of knowledge in medical and related sciences, thereby supplementing those foundations in Italy and the Federal Republic of Germany. Grants are also provided for selected research projects, scientific symposia and publication of scientific proceedings. Amongst its virological interests is included the exclusive supply of polio vaccine to Unicef for the WHO's expanded programme of immunization. Also, vaccine for the prevention of measles is supplied to the developing nations, and the Corporation markets vaccines against canine and feline viral diseases such as infectious parvovirus. On the R & D front it has been quick to realize the enormous implications of molecular biology in creating therapeutic agents against many diseases.

Of the numerous other companies in the field of human and animal health, the following deserve some mention for their viral-related activities: Beecham Pharmaceuticals (antiviral chemotherapy); Cetus (commercial application of biotechnology, in particular recombinant DNA techniques and immunological characterization of viral antigens), and the Upjohn Company (unspecified experimental science programmes in virology). Other biotechnology companies specializing in recombinant DNA technology, immunology and molecular genetics include Eli Lilly & Co.; Hoffman-La Roche; Biogen Inc.; Celltech Ltd; Merck Frosst Laboratories; Searle Research and Development Ltd; Unilever Research; and Wyeth Laboratories Inc.

Another important role of the commercial sector is the supply of biological products, reagents and equipment to the hospital, research and teaching laboratories engaged in virological activities. There are many such firms, some specializing in a very limited range of products such as accessories for the electron microscope, others providing a whole range.

The three leading international companies in this area are Flow, Gibco Europe and Wellcome Reagents, all providing a very similar spectrum of media, sera, immunologicals, etc. Their annually produced catalogues, as well as containing a list of their products, contain some very useful technical information such as procedures for virus isolation, cell culture and quality control. In addition Flow produces an information bulletin called *Nucleus*, which gives news of products, media and equipment.

Table 2.3 lists the main suppliers in the UK and gives an indication of their range of products.

2.6 NATIONAL ORGANIZATIONS — UK (*see* directory entries 405–445)

Government Bodies

The UK has a long history of both direct and indirect government involvement in virus research. In the main, four bodies — the Agricultural Research Council, the Ministry of Agriculture, Fisheries and Food, the Medical Research Council and the Public Health Laboratory Service — are of prime concern to the virologist.

Agricultural Research Council

The Agricultural Research Council (ARC) plans and coordinates research into all aspects of agriculture: soil research, fertilizers, crop husbandry, animal nutrition and, importantly from virology's point of view, animal diseases. Set up in 1931 and largely financed by the state, the Council supervises and controls about 50 research institutes as well as funding research in universities and independent research bodies.

Within the ARC the work of the following research institutes has particular relevance to virology, in terms both of the work they undertake and the publications they issue:

1. The Animal Virus Research Institute at Pirbright. The Institute has departments of Biochemistry, Epidemiology, Experimental Pathology, Genetics, Vaccine Research and Virology, which are all concerned with various aspects of foot-and-mouth disease. In addition some work on the viral vesicular diseases and rabies is conducted.

2. The Animal Disease Research Association at the Moredun Institute, Edinburgh. The Association carries out epidemiological research on scrapie and more applied research on the various diseases that attack sheep, including herpes, orf and enteric viruses.

3. The Institute for Research on Animal Diseases at Compton. The Institute has

Table 2.3 Main UK-based companies supplying virology laboratories

	BDH	Boehringer	Dynatech	Emscope	Difco	Flow	Gibco Europe	Nordic	Oxoid	Sigma	Tissue Culture Services	Wellcome
Biochemical and organic compounds	✓	✓			✓				✓	✓		
Cell cultures						✓	✓				✓	
Viral diagnostic kits						✓	✓					✓
Immunological reagents		✓	✓		✓	✓	✓	✓		✓	✓	✓
Materials and equipment for electron microscopy				✓								
Plastics for tissue culture			✓			✓	✓					
Sera/media for tissue culture					✓	✓	✓		✓		✓	✓

departments of Biochemistry, Cell Pathology and Microbiology engaged in viral research.
4. The Houghton Poultry Research Station. This research station is known particularly for its work on two virus diseases that affect chickens and turkeys — leukosis and Marek's disease.
5. The John Innes Institute, Norwich. Plant diseases are the chief concern of the Institute; typically, work is conducted on the biochemical and biophysical aspects of plant viruses. Ultrastructural studies, using the electron microscope, are also a feature of the Institute's work.
6. The Scottish Horticultural Research Institute near Dundee. Within the Department of Virology research is conducted into the biological behaviour and replication of plant viruses, particularly those of wild grasses.
7. The East Malling Research Station, Maidstone, the National Vegetable Research Station, Wellsbourne and Rothamsted Experimental Station at Harpenden are engaged in relevant research in respect to plant pathology, viral diseases of fruit, chemotherapy of vegetable virus diseases and fungal viruses. Rothamsted is additionally concerned with viruses of tropical plants.

Ministry of Agriculture, Fisheries and Food

The work of the Ministry of Agriculture, Fisheries and Food (MAFF) is closely related to, and at times overlaps with, the work of the ARC. Most of the research takes place under the aegis of the Agricultural Development and Advisory Service (ADAS). Established in 1971, ADAS exists to serve all sectors of the agricultural industry, providing in particular specialist advice on: animal diseases and their prevention; control of animal and plant pests; diseases, control and improvement of horticultural and plantation crops.

The work of four laboratories deserves special mention because of its relevance to virology. The Central Veterinary Laboratory at Weybridge undertakes a wide range of research into animal diseases as well as diagnostic and service work. The production of vaccines and other biological products used in animal health schemes also takes place at the Laboratory. The Biological Products and Standards department, which is a national control authority for veterinary products, also acts as a WHO International Laboratory for biological standards. A Rabies unit has recently been installed.

The Plant Pathology Laboratory at Harpenden is concerned with plant virus diseases and their vectors. It produces the journal *Plant Pathology* and many technical bulletins.

The Lasswade Laboratory and the Veterinary Research Laboratories, Belfast, Northern Ireland (in fact under the control of the Department of Agriculture) are mainly involved in poultry virus disease research.

Twenty-four regional Veterinary Investigation Centres, at which routine virus isolation and serology testing are carried out, complete the MAFF picture.

Medical Research Council

The Medical Research Council (MRC) was set up in 1913. At its heart lies the National Institute for Medical Research at Hampstead, which itself was established in 1922 to mount a programme into 'the filterable viruses'.

The MRC comprises four boards: Cell Biology and Disorders, Physiological Systems and Disorders, Tropical Medicine Research Board and Mental Health (the last-mentioned carries out no virus work).

The Board of Cell Biology and Disorders is funding particularly relevant work on interferon at the Laboratory of Molecular Biology (in association with Warwick University). Relevant too is the cancer research programme (in liaison with WHO) and the immunological work, using viruses as model systems, carried out at the National Institute for Medical Research.

Most of the virus work conducted by the Physiological Systems and Disorders Board takes place under the general heading of 'Infections'. At the Department of Pathology at Cambridge nucleotides of influenza A are being sequenced. MRC grants have also been awarded to the Departments of Medical Microbiology at the London School of Hygiene and Tropical Medicine and to the Royal Free Hospital's Department of Medicine, to carry out work on hepatitis A and B viruses. Work on in-utero infections by cytomegalovirus has been taking place at Guy's Hospital in London. In conjunction with the Public Health Laboratory Service at Porton, research is carried out into a particularly lethal zoonotic virus of monkeys, called herpes B.

Much of the research taking place under the Tropical Medical Research Board is concerned with yellow fever and is usually co-financed by the Overseas Development Administration of the Foreign and Commonwealth Office.

MRC collaborates with a number of official European and international bodies: these include the EEC Committee on Medical and Public Health Research, the European Science Foundation, the European Medical Research Councils, the European Molecular Biology Laboratory and the WHO's International Agency for Research on Cancer.

In addition to funding research in independent establishments the MRC also directly supervises the work of 60 research establishments, of which the following are involved to some degree in virus research: the National Institute for Medical Research; the Clinical Research Centre, Harrow; the Cellular Immunology Unit, Oxford; the Epidemiology Unit, Cardiff; the Laboratory of Molecular Biology, Cambridge; the Virology Unit, Institute of Virology, Glasgow (*see* the University of Glasgow); and the Leukaemia Unit, Royal Postgraduate Medical School, London.

Public Health Laboratory Service

The Public Health Laboratory Service (PHLS) comes under the auspices of the Department of Health and Social Security and is responsible for the prevention and control of infectious diseases via the community physicians. It provides a network of bacteriological and virological laboratories throughout England and Wales.

Its three main centres are the Central Public Health Laboratory (CPHL) at Colindale, London; the Centre for Applied Microbiology and Research (CAMR); and the Communicable Disease Surveillance Centre, also at Colindale.

The CPHL is split into a number of divisions, of which about half are engaged in some form of virus research. Studies on influenza, rubella, measles and polio are undertaken at the Epidemiology Research Laboratory, while phage typing techniques are utilized in the Division of Enteric Pathogens. The Virus Reference Laboratories conduct research in hepatitis B, entero, herpes, influenza, parvo, polyoma and rabies viruses. Antigens and sera are prepared in the Division of Microbiological Reagents and Quality Control.

The recently established CAMR at Porton is noted for its work on some of the deadliest viruses known to man. Of the 10 laboratories at the CAMR just three are of concern to the virologist. The Genetic Manipulation Laboratory works on polyoma and hepatitis B viruses as model systems for bioengineering. The facilities of the Special Pathogen Reference Laboratory — another relevant laboratory — are second to none in the UK, mainly owing to the necessity of preventing contamination of the environment by Lassa, Marbola and Ebola viruses. Attempts at producing vaccines against herpes simplex viruses and cytomegalovirus are carried out by the Vaccine Research and Production Laboratory.

The Communicable Disease Surveillance Centre acts as a monitoring and information centre for communicable diseases and provides assistance in the investigation and control of diseases such as infectious hepatitis, influenza, poliomyelitis, rubella and smallpox. It publishes regular reports on the incidence of disease.

In addition to the three main centres already outlined there are a number of Reference Laboratories (largely concerned with bacteria), 11 Regional Laboratories and 42 Public Health Laboratories in major towns and cities. They serve both as the microbiology laboratories for the hospital in which they are situated and as the centre for microbiological investigations concerning the diagnosis, prevention and control of infectious disease in the surrounding community. Much of their recent work has been associated with the campaign to vaccinate women of a child-bearing age against rubella. Other work of a virological nature includes the identification and isolation of microorganisms, and the detection of antibodies against disease.

Natural Environment Research Council

One would not immediately associate virus research with the Natural Environment Research Council (NERC) as it does not appear to fall within that organization's subject responsibilities: solid earth; inland waters; terrestrial communities; the seas, and the atmosphere. Notwithstanding this the NERC does have an Institute of Virology (formerly the Unit of Invertebrate Virology) at Oxford. The central theme of research work at the Institute is insect viruses, especially those of important pests, regardless of geographical origin. It has developed a particular interest in controlling insect pests with viruses. Early research seems to indicate that Baculoviruses could be very effective in this role, as they appear to be confined solely to invertebrates.

The viruses of wild birds, trees and grasses are also investigated, but in conjunction with the Institute of Terrestrial Ecology, another NERC body.

National Institute for Biological Standards and Control

The need for standardization and control of therapeutic products of biological research led to the setting up of the National Institute for Biological Standards and Control (NIBSC) in 1972 to coordinate this work. The Institute was the result of a merger of two divisions of the MRC's National Institute for Medical Research: Biological Standards and Immunological Products Control. In 1976 NIBSC became independent of MRC in recognition of the fact that it had statutory work to do under the Medicines Act (1968) as well as medical research. The NIBSC is now under the authority of the National Biological Standards Board.

The Institute comprises 10 departments, of which those of Viral Products and Immunology are most relevant to virology. The work of these departments is mainly applied and largely concerned with: live and inactivated viral vaccines, such as rubella and influenza; elucidating the genetic basis of virus virulence and attenuation; developing tests for the calibration of the antigen content of viral vaccines; setting up standards and reference preparations; and ensuring the quality and efficacy of manufacturers' vaccines.

The Institute is involved closely with the Department of Health's Committee on the Safety of Medicines, and the British and European Pharmacopoeia Commissions in devising, evaluating and formulating written standards and recommendations.

Professional Associations

Society for General Microbiology

Founded in 1945 and now with a membership of over 3000, the Society for General Microbiology (SGM) is the premier society for microbiology — and for that matter virology — in the UK. The Society is affiliated to both the Federation of European Microbiological Societies and the International Union of Microbiological Societies. The Society conducts its main work through eight specialist subject groups, two of which are directly relevant to virologists: the Clinical Virology Group and the Virus Group. However, the work of the other groups — Cell Biology; Cell Surfaces and Membranes; Ecology; Fermentation; Genetics; Pathogenicity; Systematics, and Teaching—may be of occasional interest as they sometimes discuss topics of a virological nature. In addition there are regional branches for Ireland and Scotland.

The principal activities of the SGM are organizing conferences and publishing. In respect of the former the main feature of the programme is the Ordinary Meeting, held 3-4 times a year. Each meeting is usually held in conjunction with an academic or professional institution and is concerned with such themes as Functional Analysis of Virus Genomes (Leeds, 1981). Whilst not all conferences are concerned with virology there is usually at least one a year devoted to it. The virology groups may contribute to the general meeting or may well hold their own meetings. The main annual symposium is a two-day affair and such is its success that it seems likely that in future two will be held a year. And significantly, as a recognition of virology's growing status, the second may become a completely virological affair. The proceedings of all meetings are generally published.

Another important event in the meetings calendar of the SGM is the annual Fleming Lecture, which is a forum for the young microbiologist who has contributed most to this branch of science in that particular year. The Lecture was inaugurated in 1975 and was given by a virologist for the first time in 1980.

As part of its active publishing programme the Society issues three journals: a 'house journal', the *Society for General Microbiology Quarterly*, which contains abstracts of papers read at conferences, news of future meetings, books reviews and reports on subjects of topical interest to microbiologists; and two scholarly journals presenting the results of current research, the *Journal of General Microbiology* and the *Journal of General Virology*. The Society also issues annually a membership directory.

Universities and Colleges

Research

British universities are not the same force in virus research as American universities. (Polytechnics do not appear to figure much in virus research.) With some notable exceptions research appears to be fragmented and lacking in direction. Nevertheless, there are about 250 virus research projects in progress at around 100 universities and university colleges. Of these well over half are conducted in departments of microbiology or virology; there are seven of the latter if we include the teaching hospitals of London University. A good indication of the subject's interdisciplinarity is provided by the fact that six other types of department figure significantly in virus research, and these are (in order of importance): biochemistry, biology, botany, pathology, medicine and veterinary medicine.

It is possible selectively to identify individual university departments which contribute significantly to virus research and these are listed below.

University of London. The University consists of a number of largely autonomous colleges, many of which are renowned in their own right for virus research; together they can muster more than 50 research projects. The London School of Hygiene and Tropical Medicine is probably the most important of them, boasting at least a dozen research projects in the field. The School has been chiefly associated with work on hepatitis A and B (it is in fact a WHO Collaborating Centre on viral hepatitis), the role of interferon in the treatment of hepatitis and aspects of vaccination.

Of the University's medical schools the one at Charing Cross Hospital is active in the study of sexually transmitted virus diseases and their role in cancer, the one at King's College Hospital is concerned with the development of techniques for the detection of rubella and other viral antibodies, and the one at St George's Hospital carries out immunological studies of cytomegalovirus infection. St Thomas's Hospital, St Mary's Hospital and St Bartholomew's Hospital medical schools all possess departments of virology; the first-named specializes in rubella virus with particular emphasis on vaccination, the second in variola, monkey pox and related viruses and the third in the clinical evaluation of the effectiveness and safety of new live influenza vaccines and viral immunity. Finally, the Institute of Ophthalmology investigates the ramifications of ocular herpes simplex virus infections.

University of Glasgow. Second in importance to the University of London as a centre for virus research is the University of Glasgow. Glasgow has its own

Department of Virology, where most of the research is carried out. However, the Veterinary School and Department of Microbiology also make their contributions to the total of about 20 research projects in progress. The Department of Virology is an MRC-designated Unit of Virology and also forms part of the Institute of Virology. The work of the Institute covers molecular genetics and immunological studies of herpes simplex virus and RNA viruses, chiefly rhabdoviruses. The Veterinary School's contribution comes in the field of tumour viruses whilst the Department of Microbiology is concerned with whooping cough vaccine.

Queen's University, Belfast. Queen's University's research effort is led by the Department of Biochemistry, which has about eight or nine research projects under way at any one time. The Department of Biology and the Department of Microbiology and Immunology also make a contribution to virus research — about three projects in both cases.

Other universities. The University of Sheffield has its own Department of Virology, which specializes in cancer research. The Universities of Liverpool (Department of Medical Microbiology), Manchester (Department of Bacteriology and Virology) and Surrey (Department of Microbiology) all have fairly strong virus research programmes, with a half dozen or so projects under way. Animal viruses are the chief concern of Liverpool, vaccinia and respiratory viruses are Manchester's specialisms and Surrey's interests span fish viruses, neonatal viral diarrhoea and disinfection of viruses in water.

Research at the University of Cambridge is a truly multidisciplinary effort with the Departments of Genetics, Clinical Veterinary Medicine, Pharmacy and Biochemistry all being involved — in all this amounts to about half a dozen research projects.

Additionally the Microbiological Departments of the Universities of Birmingham, Newcastle and Reading; the Departments of Bacteriology at the Universities of Edinburgh and Bristol; the Department of Biological Sciences at the University of Warwick; the Department of Biology at Wye College; and the Microbiology Unit at Oxford University's Department of Biochemistry mount somewhat smaller virus research programmes.

Polytechnics. Pioneering the polytechnics' contribution to virology is Sunderland Polytechnic with an ambitious research programme that includes: purification of rotaviruses and the isolation of serologically active components; the occurrence of viruses in river water and sewerage effluent; and epidemiological work associated with gastroenteritis in Sunderland.

Another way of gauging the relative strength of the aforementioned university departments in the field of virus research is to determine which are the most prolific publishers. An analysis of papers published in five virology journals over the past five years (1977–81) showed that the London School of Hygiene and Tropical Medicine accounted for 21% of the papers contributed by UK universities, Cambridge 20% and Newcastle 14%. Six other universities — Glasgow, Birmingham, Belfast, Edinburgh, Warwick and Dundee — accounted for the remaining 45% with no university exceeding 8%. (The journals chosen

Table 2.4   Organizations funding virus research in UK universities

MRC	35%
Universities	31%
Cancer research campaigns	6%
Wellcome Trust	4%
World Health Organization	4%
Agricultural Research Council	2%
Others	18%

Source: *Research in British universities polytechnics and colleges*, vol. 2 (1980) (*see* bibliography entry 10)

were *Virology*, the *Journal of Virology*, the *Journal of General Virology*, *Archives of Virology* and the *Journal of Medical Virology*.)

Major funders of virus research in UK universities are, as *Table 2.4* demonstrates, the MRC and the universities themselves. An analysis of the 'others' category illustrates how wide the interest in viruses really is: the Science Research Council; the Multiple Sclerosis Society; the Wolfson Foundation; Regional Health Authorities; the Ministry of Defence; the Association of Commonwealth Universities; the Haemophilia Society; Glaxo; and various foreign goverments. These figures, based as they are on the number of projects funded and not the actual amount of financial assistance, are inevitably rough and ready. However it is not possible to calculate research support in financial terms because figures are presented for the biological sciences as a whole and no breakdown is provided for individual fields such as virology.

Study

Virology is usually met for the first time by undergraduates when studying for microbiology degrees. The treatment of virology by the universities and colleges varies from comprehensive to superficial. In the latter case it is usually an afterthought of bacteriology.

Thirty-four universities offer courses entitled microbiology, whilst only Manchester University and Warwick University specify virology in the title of their particular degree course. High-standard degree and degree-equivalent courses in microbiology are also run by 14 polytechnics and colleges of technology. In some of these virology is offered only as an option. Virology is also taught of course as part of medical and veterinary degrees.

In the majority of cases virology is met, in earnest, only at the postgraduate level, usually in the form of a specific research project undertaken to achieve an MSc, MPhil or PhD. However, a number of higher degrees by instruction do exist, most of them also consisting of a research project. These courses are listed below.

Brunel.   Applied Cell Science and Virology.   MSc/Certificate of Advanced Study.   2 years part time.   This course comprises a fairly comprehensive study of virology and cell science and includes antiviral chemotherapy, cancer research and vaccine manufacture.

Manchester. Bacteriology and Virology. MSc/Diploma. 1 year full time. Includes general virology.

Reading. Virology. MSc. 1 year full time. All aspects of virology including tissue and cell culture methods.

Heriot-Watt. Applied Microbiology. MSc. 1 year full time. Some virology.

A number of extremely good immunology courses do exist which may be of interest to virologists, such as the ones held at Brunel and Chelsea College, University of London.

Higher degrees by research are offered by many universities. The specific area in which research is to be undertaken is usually at the discretion of individual instructors and varies from year to year. Universities and colleges normally offering research projects in virology include those listed in *Table 2.5*.

Table 2.5 Some UK universities and colleges awarding higher degrees in virology

Belfast	MSc, PhD
Birmingham	MSc, PhD
Bristol	MSc, PhD
Dundee	MSc, PhD
Edinburgh	MPhil, PhD
Glasgow	MSc, PhD
Leeds	MPhil, PhD
Manchester	MSc, PhD
Newcastle	MSc, PhD
Oxford	MSc, DPhil
Reading	MPhil, PhD
Sheffield	MSc, PhD
Surrey	MPhil, PhD
Wolverhampton Polytechnic	MPhil, PhD

Other Organizations — Cancer Research Bodies

Much useful virus research has been carried out in the name of cancer research. Whilst it is true that as yet there is no conclusive link between human cancer and viruses, there are many animal examples. Viruses provide ideal models for learning about cancer processes, particularly in their relationships with cultured cells. Cancer research is by nature multidisciplinary, much of it being concerned with chemotherapy, radiobiology and immunology.

There are a number of institutions in the UK existing financially on voluntary contributions and which in turn provide grants to research centres at universities and teaching hospitals, to aid cancer research.

Cancer Research Campaign

Formerly the British Empire Cancer Campaign for Research, the CRC has its own laboratory, Gray Laboratory at Mount Vernon Hospital in Northwood, London, as well as jointly supporting (with the MRC) the Beatson Institute for Cancer Research, Glasgow; the Palilson Laboratories, Manchester, and the Institute of Cancer Research, Royal Cancer Hospital, London.

Imperial Cancer Research Fund

One of the oldest institutes, founded in 1902, the Fund has two main laboratories, at Lincoln's Inn Fields and Mill Hill (both in London) and three extramural units, at St Bartholomew's Hospital, Guy's (New Cross) Hospital and University College London. The virological nature of their work is reflected in the names of their departments: the Departments of Molecular Virology, Viral Oncology, Tumour Virus Genetics and Tumour Virology at Lincoln's Inn. At Mill Hill most of the viral work is conducted in the Department of Immunopathology.

Institute of Cancer Research

The Institute comprises the Chester Beatty Research Institute, London and the Institute of Cancer Research at Sutton, Surrey.

.7 NATIONAL ORGANIZATIONS — USA (*see* directory entries 446–451)

Government Bodies

US Department of Health and Human Services

The Center for Disease Control in Atlanta is divided into a number of Bureaux of which those of Epidemiology and Laboratories are involved in virus research. The Bureau of Epidemiology includes a Viral Disease Division, specializing in Respiratory and Special Pathogens, and a Hepatitis Laboratory Division, which uses primates for research. The Bureau of Laboratories has a Virology Division, with a specialism in zoonotic viruses, and a Vector Borne Disease Division, concerned with ecological aspects of virus diseases.

Occasionally the work of the Food and Drug Administration at Rockville is of interest to the virologist; much of this work is undertaken collaboratively with the National Institutes of Health.

National Institutes of Health

The National Institutes of Health are the principal agency of the Department of Health for biomedical research, research training and biomedical communication in the interests of public health. They are the main funder of research in the field.

Outside the universities the Institutes are the major contributor to virus research in the USA. They comprise 12 institutes, of which four are of particular interest and are centred at Bethesda, Maryland.

The National Cancer Institute, founded in 1931, broadly supports research into the causes, detection, diagnosis and treatment of cancer. In the late 1960s it began an intensive targeted and coordinated research programme on the role of viruses in cancer, with emphasis on RNA viruses. The Virus Cancer Program, as it was called, became an integral part of the National Cancer Act of 1971, which included a carcinogens testing programme. The Laboratory of Cellular and Molecular Biology conducts field studies and statistical exercises in support of the programme.

The National Institute of Allergy and Infectious Diseases (NIAID) conducts and supports research in microbiology aimed at: solving new problems in allergic diseases; unsolved problems in bacterial diseases; and, more generally, developing a useful body of knowledge on the viral diseases. In 1962 it initiated the Research Reagents Program for the purpose of producing, testing and distributing virus reagents, consisting of purified seed virus or antigens and corresponding antisera. By 1980 reagents for 12 main groups of viruses were available, as well as some interferons. The reagents are distributed to investigators throughout the world, usually free of charge.

The Program involves the commercial sector, international agencies such as the WHO and FAO — particularly in relation to the joint Program for Comparative Virology — and other US government groups such as the Food and Drug Administration, the Department of Agriculture, the National Cancer Institute and the American Type Culture Collection.

The Institute has published a *Catalog of research reagents 1978–1980* and was also responsible for the publication of a series of six reports written by the Virology NIAID Task Force (1979) which encompassed the whole spectrum of research.

Two other Institutes, the National Institute of Neurological and Communicative Disorders and Stroke and the National Institute of General Medical Sciences, are also active, albeit to a somewhat lesser extent, in the field of virology.

US Department of Agriculture

Animal and plant virus research is the concern of the Science and Education Administration's Agricultural Division. It conducts and supports research and teaching programmes in agricultural research within the area of animal and plant production. Important laboratories active in these fields include: the Beltsville Agriculture Research Center; the Plum Island Animal Disease Center, New York; the National Animal Disease Center, Ames, Iowa; the Animal Disease Research Laboratory, affiliated to Washington State University; the Regional Poultry Research Laboratory, East Lansing; and the Arthropod-borne Animal Disease Research Laboratory at Denver.

The overall objective of the Department's animal health research is to reduce costs of meat and animal products by developing methods for the diagnosis, prevention, control and eradication of diseases of animals. Additionally, broadly based research projects are conducted such as identification and classification of slow viruses and general research work on other viruses of bovine leucosis, rhinotracheitis, blue tongue, Newcastle, swine influenza and enteric viruses.

Department of Defense

Whilst not obviously associated with virus research, the Department of the Army's Regional Veterinary Consultants/Coordinators in fact run the Walter Reed Army

Medical Center in Washington, which publishes many papers on infectious viral diseases. There are also six other medical centres, the US Medical Research and Development Command at Fort Derrick, Frederick, and the US Army Academy of Health Sciences at Fort Sam Houston, some of whose work is of marginal or occasional interest to the virologist.

Professional Associations

National Academy of Sciences

The National Academy of Sciences (NAS), a private non-profit-making organization established in 1863, is concerned with the advancement of science and the benefits this can bring. It chartered the Institute of Medicine in 1976 to tackle the problems of providing adequate health services, which it does by identifying important issues related to health and medicine.

The NAS set up the National Research Council (NRC) to carry out studies by means of appointed committees of scientists. The NRC is composed of a number of assemblies of which the Assembly of Life Sciences, subdivision Biological and Medical Sciences, is of prime interest.

The NAS is divided into a number of sections to which members are assigned according to their specialism; virologists can be found under Medical Microbiology and Immunology.

It publishes the *Proceedings of the National Academy of Sciences*.

American Society for Microbiology

Founded in 1899, with a current membership of over 20 000, the American Society for Microbiology (ASM) is one of the largest and oldest microbiology societies in the world. Indeed, it is nearly seven times the size of its British counterpart, the SGM, and predates it by over half a century. The Society is affiliated to the IUMS and has branches as far afield as Mexico and Argentina.

The Society conducts much of its affairs through a number of divisions, one of which is Virology; the others are Environment and Applied, General, and Medical and Immunology.

The ASM is, by any standards, a prolific publisher, issuing eight monthly and two quarterly periodicals and an annual abstract of its meetings. The most important journal from a virological point of view is the *Journal of Virology*. The other journals include: (monthlies) *Antimicrobial Agents and Chemotherapy, Applied and Environmental Microbiology, Infection and Immunity,* the *Journal of Bacteriology,* the *Journal of Clinical Microbiology, Molecular and Cellular Biology,* and *ASM News,* the Society's mouthpiece; and (quarterly) *Microbiology Reviews. Abstracts of Annual Meetings of the American Society for Microbiology* is the aforementioned service abstracting conference papers; it appears once a year.

Another annual of a similar but more narrative style is the *Microbiology* series, which extends back to 1975, when the first volume, *Microbiology — 1974* appeared. *Microbiology — 1980* contains material from ASM conferences and symposia that merited particular attention. The virology content varies, *Microbiology — 1980* being a particularly good year. Of all the meetings the ASM organizes, the Annual Meeting is the most important. The 81st, a week-long affair that took place in Dallas in 1981, was divided into colloquia, symposia, seminars and round tables; virology-related topics figured prominently.

Two ASM books of great importance and popularity are *Manual of clinical microbiology* (3rd edn, 1980) and *Manual of clinical immunology* (2nd edn, 1980). These have been of particular use in hospitals as well as in research and teaching laboratories. To supplement these and other practical manuals, the ASM has produced brief descriptive pamphlets on a variety of technical and procedural topics entitled *Cumitechs*. *Cumitech 15: Laboratory diagnosis of viral infections* (1982) is the most relevant of the series.

In the field of education, the Board of Education and Training of the ASM publishes a regular directory of microbiology courses in the USA. The most recent is the *1980 Directory of colleges and universities granting degrees in microbiology* (*see* p. 24). In addition to the directory and many other publications, the Board produces audiovisual facilities for teachers and lecturers in many microbiological areas. A catalogue lists the entire collection of slides: *1982 Slide Catalog*.

The Society also awards annual cash prizes for outstanding contributions to research, teaching and study in the field of microbiology.

Universities and Colleges

The contribution made by the universities of the USA to the rapid growth and development of virology cannot be overemphasized. As in many other branches of science the work of the Americans has been prolific, partly because of the sheer numbers of research workers in the field and partly because of the greater resources at their disposal. Their domination of the periodical literature, including the highly prestigious journals, must also testify to the quality of much of their research.

It is not easy to gauge the financial support given to virus research in the US because the various funding organizations do not provide a detailed enough subject breakdown in their accounts. A rough idea, however, can be obtained by enumerating the acknowledgements for financial assistance made at the end of the papers submitted to such core journals as *Virology* and the *Journal of Virology*. Whilst we accept that this may be an imprecise measure, *Table 2.6* does give some idea of research funding in virology for 1980. This table shows that government finance amounts to 86% in total and that of the major private funder — the American Cancer Society — to only 10%.

It is not possible to describe comprehensively the activities of all the universities where virus research is undertaken because of their sheer number; in fact no up-to-date guide to research in American universities and colleges exists. Without such an information source it was decided to examine the major virology journals and guides to dissertations to obtain some insight as to what is being carried out and where. From such sources it was possible to identify over 50 universities conducting virus research as evidenced by journal articles, theses and reviews. From this number we selected just over 30 universities to discuss in a little more detail. First is listed the name of the university and the interested department(s), followed by an indication of its subject specialisms.

University of Alabama: Department of Microbiology, Medical Center, Birmingham, Alabama

Table 2.6 Sources of US research funding in virology

US Public Health Service	27%
National Institutes of Health	26%
American Cancer Institute	20%
American Cancer Society	10%
National Science Foundation	6%
National Institute for Allergy and Infectious Diseases	4%
Department of Energy	3%
Hartford Foundation, Welch Foundation, Helen May Whitney Foundation	1%
Others	3%

    Carp virus: analysis of phosphoproteins
    Rhabdovirus: mapping and transcription
    Vesicular stomatitis virus: sequence analysis
    Influenza C virus: structural analysis
    Bunyaviruses: genetics

Baylor College of Medicine: Department of Virology and Epidemiology and Department of Cell Biology, Houston, Texas

    Herpes simplex virus: polypeptide characterization and biochemistry of transformation
    Rotavirus: infectivity
    Tumour viruses: oncogenesis

California Institute of Technology: Department of Chemistry, Pasadena, California

    Feline leukaemia viruses: biochemistry
    Bacteriophage T_4: infectivity

University of California: Department of Molecular Biology and Virus Laboratory, Berkeley; Department of Molecular Biology and Biochemistry, Irvine; Department of Microbiology and Immunology, UCLA School of Medicine, Los Angeles; Department of Microbiology and Parvin Center Research Laboratory, Molecular Biology Institute, San Francisco, California

    Paramyxovirus: protein analysis of nucleocapsid
    Avian retroviruses: genetic analysis
    Visna maedi: DNA structure
    Bacteriophage: replication and transcriptional mapping of genome
    Herpes simplex virus: replication
    Human and simian adenoviruses
    Turnip crinkle virus: RNA studies
    Cauliflower mosaic virus: structure of genome
    Murine leukaemia virus: phosphoprotein analysis

University of Chicago: Marjorie B. Kovler Viral Oncology Laboratory, Department of Medicine and Committee on Virology, Chicago, Illinois

> Herpes simplex virus: phosphoproteins and DNA sequencing
> Retroviruses: genetics
> Epstein–Barr virus: general studies

Albert Einstein College of Medicine, Bronx, New York

> Hepatitis B virus: transcription of gene sequences

Frederick Cancer Research Center: Biological Carcinogenesis Program, Frederick, Maryland

> Oncogenic viruses: in particular, studies on mouse mammary tumour

University of Florida: Department of Preventative Medicine, College of Veterinary Medicine and College of Medicine, and Department of Plant Pathology, Gainesville, Florida

> Adeno-associated viruses: latent infection
> *E. coli* T_5: infection
> Squash mosaic virus: structural gene analysis
> Goat retroviruses: morphology and immunology

Harvard Medical School: Department of Microbiology and Molecular Genetics, Boston, Massachusetts

> Murine retroviruses: characterization of genome
> Herpes simplex virus: molecular genetics

University of Illinois at the Medical Center: Department of Microbiology, Chicago, Illinois

> SV_{40}: transformation

University of Iowa: Department of Microbiology, College of Medicine, Iowa City, Iowa

> Cytomegalovirus: replication

Johns Hopkins University: Oncology Center and Department of Pharmacology and Experimental Therapeutics, School of Medicine, Baltimore, Maryland

> Mouse and human cytomegalovirus: chemotherapy and genetics of structural and non-structural proteins, replication
> Retroviruses: molecular genetics
> Epstein–Barr virus: genomic organization

University of Maryland: Department of Botany

> Cucumber mosaic virus: associated RNA

Massachusetts Institute of Technology: Center for Cancer Research and Department of Biology, Cambridge, Massachusetts

> Vesicular stomatitis virus: replication, virion assembly
> Murine leukaemia virus: mutants and biochemical analysis of polypeptides, cell culture studies

Adenoviruses: general studies

Michigan State University: Department of Biological Chemistry, East Lansing, Michigan
Polyoma virus: replication

University of Minnesota, St Paul, Minnesota
$E.\ coli$ T_5 phage: infection studies

University of Missouri: Division of Biological Sciences, Columbia, Missouri
Retroviruses: molecular biology

New York University School of Medicine: Department of Pathology, New York
SV_{40}: protein chemistry

College of Medicine and Chemistry of New Jersey: Department of Microbiology, New Jersey
Sindbis virus: replication in insect cell culture
Influenza, Semlike Forest, Sendai: general
Human papovaviruses: genome analysis studies

Pennsylvania State University: Department of Microbiology, Cell Biology, Biochemistry and Biophysics and Cancer Research Center at Milton S. Hershe Medical Center, Philadelphia, Pennsylvania
Phage: genetics
Herpes simplex virus: replication, infection and biochemistry of transformation; characterization of DNA
Human cytomegalovirus: immunoprecipitation of polypeptides

University of Rochester, School of Medicine and Dentistry, Rochester, New York
Phage T_7: genetic analysis of DNA and replication
Mycoplasmaviruses: DNA synthesis

Rockefeller University, New York
Adenoviruses: transcription
Avian retroviruses: recombination
Paramyxoviruses: inhibition of neuraminidase

The Salk Institute, Tumor Virology Laboratory, San Diego, California
Murine leukaemia virus: integration
Rous sarcoma virus: temperature-sensitive mutants
Semlike Forest virus: characterization of proteins
Retroviruses: genomic organization

St Louis University Medical Center, Institute for Molecular Virology, St Louis, Missouri
Retroviruses: molecular biology

University of Southern California: Department of Microbiology and Neurology, School of Medicine, Los Angeles, California

Reticuloendotheliosis viruses: gene sequencing and oncology
Mouse hepatitis virus: structure of RNA
Avian sarcoma viruses: transformation
Avian erythroblastosis virus: transformation

University of Texas: Department of Microbiology, Health Science Center, Dallas, and Department of Tumor Virology, M. D. Anderson Hospital and Tumor Institute, Houston, Texas
Avian sarcoma virus: replication
Murine leukaemia virus: replication

University of Utah Medical School: Departments of Cellular, Viral and Molecular Biology and Biochemistry, Salt Lake City, Utah
Poliovirus: replication
Vesicular stomatitis virus: ribonucleoprotein studies

University of Wisconsin: McArdle Laboratory for Cancer Research, Madison, Wisconsin
SV_{40}: molecular genetics
Avian leukosis virus: cytopathic effects
Avian retroviruses: studies on DNA polymerase and ribonucleotides
Vaccinia virus: enzyme synthesis

University of Wisconsin — Madison, Madison, Wisconsin
Polioviruses: coat protein analysis
Bacteriophage lambda: restriction mapping of phages

Wistar Institute of Anatomy and Biology, Philadelphia, Pennsylvania
Murine sarcoma viruses: molecular genetics

Vanderbilt University: School of Medicine, Nashville, Tennessee
Herpesviruses: DNA replication
Vesicular stomatitis virus: inhibition of replication

Yale University: Department of Epidemiology and Public Health, and Yale Arbovirus Research Unit, Yale University School of Medicine, New Haven, Connecticut
Colorado tick fever: genetics
Bunyavirus: genetics

2.8 NATIONAL ORGANIZATIONS — OTHER COUNTRIES

(*see* directory entries 452–478)

Australia

The federal government directly funds and carries out research through the National Health, Medical Research Council and the Commonwealth Scientific and Industrial Research Organization (CSIRO). It indirectly supports research through general grants to universities and special departmental institutions and laboratories.

Accounting for 40% of all aid in the field of biology, CSIRO issues, annually, a publication called *CSIRO Research Programs*, which contains descriptions of the 800 or so research programmes it supports. By far the most important CSIRO body for virology is the Animal Health Division, which conducts research into the causes, diagnosis, control and eradication of economically important diseases of livestock such as blue tongue virus and other arboviruses. The Arbovirus Research Program was originally established to isolate and identify endemic arboviruses and to study their pathogenicity and epizootiology. The Program was however expanded to examine exotic viruses such as akabane virus, that are potential threats to Australia. On completion of the Australian National Animal Health Laboratory in Geelong — currently under construction — research into high-security microbiological diseases will be greatly facilitated. CSIRO will also provide laboratories for foot-and-mouth disease vaccine, and quality control work on other viral vaccines.

The National Biological Standards Laboratory at Canberra, which is part of the Commonwealth Department of Health, implements government policies and administers Commonwealth legislation on health and related fields. From a virological standpoint it is a control authority on medical and veterinary products such as viral vaccines.

The Institute of Medical and Veterinary Research in Adelaide performs research into diseases of animals and humans and also provides a diagnostic pathology service.

Universities active in the field of virus research and education are: the Australian National University at Canberra, whose affiliated John Curtin School of Medical Research has made major contributions to influenza research (a Virus Ecology Unit is also located there); and the University of Queensland, where virus research is carried out in a number of different departments — Agriculture, Veterinary Pathology and Public Health, and Microbiology.

Canada

Canada's contribution to virus research is dominated by its universities. Direct government involvement is largely confined to the Department of Agriculture, although the government does provide, via such bodies as the Medical Research Council, much of the funding that supports virus research in universities and hospitals.

Most prominent of all the academic establishments is the Institut Armand-Frappier, which although part of the University of Quebec is in fact 150 miles away from Montreal. The Institut is a centre of postgraduate study and has an extremely prestigious and active research programme. Its main lines of research include studies into: the fundamental properties of respiratory and herpetic viruses; methods of diagnosing viral infections; and the development of new vaccines and chemotherapeutic substances. The Institut also awards postgraduate degrees in virology.

The University of Toronto and McGill University also deserve mention for their virological studies, as does the University of Montreal, which has a Comparative Virology Research Group in the Faculty of Medicine.

France

Research is carried out partly by the Ministry of Health and Social Security, the Institut National de la Santé et de la Recherche Médicale and the Institut Pasteur, partly by universities and partly by the Ministry for Universities at the Centre National de la Recherche Scientifique (CNRS).

The CNRS is the body that proposes to the government the means of doing research and how to allocate funds. Its committee forms policy and controls 141 laboratories and research centres employing nearly 8000 workers. Laboratories where virus research takes place are: the Institut National de la Santé et de la Recherche Médicale and the Institut Pasteur, both in Paris; the Centre de Génétique des Virus in Gif-sur-Yvette; the Laboratoire Central des Recherches Vétérinaires in Maison Affore; the Institut de Recherches Scientifiques sur le Cancer, Villejuif; and the Institut de Recherche en Biologie Moléculaire at the University of Paris VII.

Federal Republic of Germany

There is no central control of research in Germany; instead, universities, clinics, research associations, private research institutes and the Max Planck Association carry out the work, sometimes with, sometimes without, government finance.

Probably the most important and widely regarded body is the Max-Planck-Gesellschaft zur Förderung der Wissenschaften EV (Max Planck Society for the Advancement of Science) based in Munich. The Society, previously in Berlin and known as the Kaiser-Wilhelm-Gesellschaft, provided the world with many of the early virologists. The Society today maintains 46 self-administering Research Institutes and 3 project groups. The Max-Planck-Institut für Virusforschung at Tübingen is the premier virus research centre in Germany. In the same location is the Bundesforschungsanstalt für Viruskrankheiten der Tiere, which includes a Rabies Section.

Other laboratories include the Institut für Virologie at Giessen, which is part of the Justus Liebig-Universität, the Institute of Medical Microbiology in the faculty of Veterinary Medicine of the Ludwig-Maximilians-Universität in Munich and the European Molecular Biology Laboratory in Heidelberg.

India

Much of today's virus work in India is directed towards epidemiological or immunological problems and particularly concerns the development of methods of prevention and control.

The Indian Council for Medical Research, founded in 1911, assists in financing medical research and maintaining a number of institutes. The most important of these is the Virus Research Centre in Poona. The Institute, established in 1952 under the joint auspices of the Council and the Rockefeller Foundation, was primarily interested in studying arthropod-borne viruses but has been frequently called upon to help in the investigation of other medical and veterinary virological problems.

The Haffkine Institute, Bombay, has an Enterovirus Research Unit, which works in the area of polio vaccines and sera. Another Enterovirus Research Unit is situated at the Pasteur Institute of Southern India, Canoor, where work is also carried out on rabies, influenza, adenovirus, and pox viruses.

The Association of Microbiologists of India promotes the study and application of microbiology in India. It also provides a forum for discussion through its conferences and its quarterly journal, the *Indian Journal of Microbiology*.

Relevant research in the university sector can be found at the Tamil Nadu Agricultural University in the Microbiology and Plant Pathology divisions and at the Assam Agricultural University, College of Veterinary Science.

Japan

Nihon Gakujutsu Kaigi (the Science Council of Japan) was founded in 1949 as the government-organized representative of all Japanese scientists. Its functions are to promote and reflect scientific development throughout national life, industry and commerce, to coordinate scientific research and to liaise with scientific organizations abroad.

Japan, significantly, is the only country to have a society devoted to virology: Nippon Virusu Gakkai (the Society of Japanese Virologists), which publishes *Virus*, a little-known but long-established journal of virology, as well as *Microbiology and Immunology*. These periodicals are published in Japanese with summaries in English.

Nippon Shokubutsu-Byori Gakkai (the Phytopathological Society of Japan) looks after the interests of plant virologists as well as of those interested in other plant diseases. Most of the research in this area is conducted at the National Institute of Agricultural Sciences in Tokyo.

The National Institute of Animal Health is concerned with viruses of veterinary importance and is composed of a central laboratory, four regional centres and a poultry disease centre. The Institute, run by the Ministry of Agriculture and Commerce, is particularly interested in swine vesicular disease. It publishes the *NIAH Quarterly*, which often contains a number of high-standard virology papers. Also carrying out veterinary research, the Department of Virology at Tokyo is noted for its work on the oncogenic viruses. It publishes the *Kitazato Archives of Experimental Medicine*. The Institute is affiliated to the Kitazato University, whose Department of Microbiology in the School of Veterinary Medicine and Animal Sciences conducts similar research.

One of the main centres for medical research is the National Institute of Health in Tokyo. Its Department of Virology and Rickettsiology and Department of Enteroviruses are both WHO collaborating centres.

Japan has one of the most active commercial sectors in virus research. Two companies with extensive research facilities are Shionogi Research Laboratories in Shiga and the Eisai Company Ltd in Tokyo.

USSR

Most learned societies and research institutes are attached to relevant Academies and Ministries in the USSR. Altogether there are five sections of the Academy of Sciences. Section II, Chemistry and Biological Sciences, contains the Department of Biochemistry, Biophysics and Chemistry of Physiologically Active Compounds, attached to which is the Institute of Microbiology and All-Union

Microbiology Society. The Academy of Medical Sciences oversees the Institute of Poliomyelitis and Virus Encephalitis and the D. I. Ivanovsky Institute of Virology. Other important laboratories for virology research are:

Moscow Institute of Virus Preparations
All-Union Influenza Research Institute
L. Pasteur Institute of Epidemiology and Microbiology
A. Kinshtenstein Institute of Microbiology (Latvian Academy of Sciences)
Oncology Research Centre
Institute of Molecular Genetics
N. F. Gamaleya Institute of Epidemiology and Microbiology
Institute of Experimental Biology
Leningrad Institute of Experimental Medicine
Laboratory of Experimental Immunobiology

3 Conferences

3.1	The role of conferences	63
3.2	Guides to forthcoming conferences	64
3.3	The major virology conferences	65

3.1 THE ROLE OF CONFERENCES

It has been said of conferences that they of all information channels offer the greatest range, both in degree and number, of opportunities for scientific communication. So it is perhaps not surprising that conferences, be they called congresses, symposia, meetings, seminars or workshops, are on the increase in virology. Thus in the first half of 1981 alone 10 conferences of special relevance to virologists took place throughout the world; they were:

- International Symposium on Viral Hepatitis
- British Society for Antimicrobial Chemotherapy
- Genetics as a Tool in Microbiology
- Immunology of Virus Infections
- International Conference on the Biology of Interferon
- 3rd International Symposium on Rapid Methods and Automation in Microbiology
- 12th International Congress of Chemotherapy
- Epidemiology of Plant Virus Diseases
- 5th International Congress of Virology
- Molecular and Cellular Aspects of Microbial Evolution

With such a busy agenda the keen virologist would hardly be seen at his or her laboratory. It is for this reason that the burden of conference attendance is usually shared out evenly amongst all senior members of staff.

Conferences in virology perform at least three important functions. First, they provide an opportunity to obtain first-hand details of the latest work in a relatively painless and perhaps enjoyable way. However, whilst some speakers use conferences to report interim research results, many more repeat information already previously published. In the case of papers delivered by foreign speakers, particularly those from the Eastern bloc, this may not be quite so serious as it might seem, because language and journal access problems might mean that the conference is the delegates' first acquaintance with the work. Secondly, conferences provide the opportunity to question the speaker and thus obtain a better understanding of the work. Furthermore this understanding can be greatly enhanced by the informal, off-the-cuff remarks obtained in personal discussions outside the formal conference sessions. Finally, and most importantly, conferences are also social gatherings through which contacts can be made or renewed. Personal contacts can of course unlock information channels that would otherwise remain locked, and the more extensive one's range of contacts of course, then the better one's personal information network is — and as we have hinted already, it is through this network that scientists obtain most of their information. It should be remembered that much of the informal exchange of ideas and information that is conducted through these social gatherings may have little at all to do with the topics dealt with in the conference; the conference may well be an excuse to enable this exchange to take place. From all of this it can be seen that the published proceedings (*see* sections 4.4 and 5.6) can never furnish a true record of all that went on at a conference.

3.2 GUIDES TO FORTHCOMING CONFERENCES (*see* bibliography entries 16–18)

Once a researcher becomes an established member of the scientific community he or she will have little difficulty in keeping informed as to forthcoming conferences; in many cases he will be automatically circulated with details (especially if he is a regular attender), sometimes asked to attend or even requested to give a paper. Those less fortunate can obtain their intelligence (normally about 6–9 months prior to the event) relatively straightforwardly by simply scanning the advertisements or notes section of the relevant journals, most notably:

Society for General Microbiology Quarterly
ASM News Monthly
FEMS
Intervirology
Bulletin d'Information/Newsletter
Phytopathology
Zentralblatt für Bakteriologie, Mikrobiologie und Hygiene
Soviet Progress in Virology

The larger and most important conferences will also be advertised in *Nature, Science, New Scientist, Bioscience, Lancet, New England Journal of Medicine, Biologist, Cancer Research* and *Veterinary Record*.

For those requiring a more comprehensive picture, those weary of relying upon the fruits of serendipity alone and those seeking conferences outside their particular field of expertise there are specialist tools, albeit published at the general level of science rather than virology, that are solely devoted to the listing of forthcoming meetings. Three such lists deserve special mention. *Forthcoming International Scientific and Technical Conferences* is a quarterly journal published by Aslib. Details provided for each conference include date, title, location and organizer/contact. To find virology conferences one needs to use the subject index. However, as it is based on keywords in the conference title meetings can be found only if the term under which one is searching is in the title, so it pays to search under a number of virological terms (e.g. *virus, virological, viral* and under the specific names of virus diseases). Readers should be warned that this can be a long and somewhat imprecise job — it is impossible to be really certain that you have found them all. There is also an alphabetical index to geographical locations and organizations. Searching under the latter for known virological organizations can improve the precision of the subject search. In the most recent year surveyed (1981), 10 virological meetings were publicized in all, with the period of notice varying between 6 and 12 months.

Possibly the most elaborate and comprehensive listings are offered by the two sister publications: *World Meetings: outside the United States and Canada* and *World Meetings: United States and Canada*. Published by the Technical Meetings Information Service, the two quarterlies provide a detailed two-year guide to future medical, scientific and technical meetings. Both have exactly the same format, being divided into two sections: (1) date section and (2) index section. The first section arranges conferences chronologically (in quarters) and provides the following additional details for each: name, headquarters, location, sponsor, contact, technical content of meeting, estimated attendance and any restriction thereon, availability of abstracts or papers, exhibitions, deadlines for submissions. The index section provides access via keyword (this presents the same problem to the virologist as the index of *Forthcoming International Scientific and Technical Conferences*), date of meeting, location, sponsor, and deadline for paper or abstract. The 1980 volume surprisingly listed no virology conferences. A few are listed in volumes prior to the year in which the conference takes place, though most are notified 2–4 months in advance. Unfortunately, because each monthly issue is extremely late in arriving (5 months is common, certainly as far as the UK is concerned), some conferences are actually under way when notification is received.

3.3 THE MAJOR VIROLOGY CONFERENCES (*see* bibliography entries 19–31)

Because of the number and diversity of conferences on virus research, it would be virtually impossible to list them all, particularly the 'one-off' conferences. Instead an account will be given of all the main, regularly occurring conferences of relevance to the virologist. Incidentally there is a growing tendency to base a textbook or monograph on the proceedings of a conference. Where this has occurred the book has been dealt with elsewhere, in the book section. A reference is provided in such cases.

Annual Meetings of the American Society for Microbiology

Beginning in 1900, this amounts, certainly, to the longest-running series of meetings held in microbiology. The 81st, held in Dallas in 1981, was divided, as usual, into colloquia, which were mainly of a bacterial nature; symposia, which discussed the advances in pathobiology of virus infections, amongst other topics; seminars, of which two were devoted to DNA viruses and RNA viruses; and round tables, which were concerned with a number of administrative and educational subjects. *Abstracts of the Annual Meetings of the American Society for Microbiology* and the *Microbiology* series contain the published proceedings.

Cold Spring Harbor Conferences on Cell Proliferation

This conference series is a companion to the Cold Spring Harbor Symposia on Quantitative Biology, but is a more recent innovation, having started in 1974. The subject material may overlap; for example, Viruses in Naturally Occurring Cancers was published in 1980 as a two-book set. Topics included herpesvirus-associated tumours of man, hepatitis B virus and hepatocellular carcinoma and candidate human retroviruses and papovaviruses associated with neoplastic diseases of animals. Other relevant published proceedings include:

Vol. 5	Differentiation of Normal and Neoplastic Hematopoietic cells (1978)
Vol. 4	Origins of Human Cancer (1977)

Cold Spring Harbor Symposia on Quantitative Biology

The internationally renowned Cold Spring Harbor Laboratory began its Symposia on Quantitative Biology nearly fifty years ago. Whether on specific viral topics or not, the symposia are of prime importance to virologists and all biologists. In 1979 Viral Oncogenes were discussed. The proceedings were published in 1980 as a two-volume set. The 141 papers covered all aspects of the molecular biology of both DNA and RNA viruses.

Earlier symposia of interest include:

XVIII	Viruses (1953)
XXXII	Antibodies (1967)
XXXIII	Replication of DNA in Microorganisms (1968)
XXXIX	Tumor Viruses (two-book set) (1974)
XLII	DNA — Replication and Recombination (two-book set) (1978)

Symposia of the European Association against Virus Diseases

The 18th Symposium of the Association was held in Stirling in 1981 and amongst the subjects discussed were enteric viruses, hepatitis, interferon and antiviral agents and virus infections in immuno-compromised hosts.

These symposia, begun in 1948 and held at most two years apart, attract the leading names in medical virology. Unfortunately their proceedings do not appear to be generally available to non-attenders.

Developments in Biological Standardization

Previously known as *Progress in Immuno-biological Standardization* and *Symposia Series in Immuno-biological Standardization*, this series contains the Proceedings of Congresses, Symposia and Meetings of the International Association of Biological Standardization (IABS), which began in 1955. Topics discussed include standardization, production and control of sera, vaccines and other similar products. The work of the IABS is intimately linked with that of the WHO, some meetings being co-organized.

Recent symposia of interest to virologists, although nearly all are indirectly relevant, include:

Vol. 39   Influenza Immunization, Geneva (1977)
Vol. 40   Standardization of Rabies Vaccine for Human Use Produced in Tissue Culture (Joint WHO/IABS), Marburg (1977)
Vol. 47   Reassessment of Inactivated Poliomyelitis Vaccine, Bitthoven (1980)
Vol. 48   Herpesviruses, Lyon (1981)

Topics discussed at the Herpesvirus Symposia included immunopathology, diagnostics, vaccines and immunotherapy. Publishers, as for all these proceedings, were S. Karger.

FEBS Proceedings of Meetings

This series, begun in 1966, contains the proceedings of meetings of the Federation of European Biochemical Societies. A number of symposia have been on virus subjects:

Vol. 10   The Biochemistry of Virus Replication (1968)
Vol. 22   Virus–Cell Interaction and Viral Antimetabolites (1972)
Vol. 27   RNA Viruses/Ribosomes (1974)
Vol. 39   Organization and Expression of the Viral Genome (1977)

FEMS Symposia

The Federation of European Microbiological Societies have held numerous meetings since their inception in 1977. As yet only one has been of direct virological interest. 'Arboviruses in the Mediterranean Countries' was the subject of the 6th FEMS Symposium held under the auspices of the Yugoslav Academy of Science and Arts at Zagreb in 1978. The proceedings were published by Gustav Fischer Verlag.

The Harvey Lectures

These annual lectures commenced in 1906 and are held under the auspices of the Harvey Society of New York, which is in turn under the patronage of the New York Academy of Medicine. The content of the lectures is usually broad in subject matter, topical and given by authorities in the relevant field. One, sometimes more, of the eight lectures given each year is in the area of virology.

Series 69   On the Origin of RNA Tumour Viruses (1975)
Series 70   The Strategy of RNA Viruses (1976)
Series 71   Slow Infections with Unconventional Viruses (1977)
Series 72   Molecular Epidemiology — Influenza as Archetype (1978)

ICN–UCLA Symposia on Molecular and Cellular Biology

While most of the annual symposia may have something in them for the virologist with molecular biology interests, the three most recent ones are relevant to all virologists: *Animal virology* was published in 1976 and considered broad issues and trends in animal virology with emphasis on the replication of tumour viruses and negative strand viruses. Curiously a chapter on plant viruses was also included. The volume was edited by D. Baltimore, A. S. Huang and C. F. Fox. *Persistent viruses* contains the proceedings of the 1978 symposium and was edited by J. G. Stevens, G. J. Todaro and C. F. Fox. It was published in the same year. *Animal virus genetics* was published as the proceedings of the 1980 symposium. It focused on basic genetic models, the use of genetic approaches to study basic problems in molecular biology, and the study of virus–host interaction using genetic systems. In addition, workshops were organized by virus groups rather than by research theme. The volume was edited by B. N. Fields, R. Jaenisch and C. F. Fox. Academic Press published the three mentioned proceedings.

International Conferences on Comparative Virology

Three conferences have so far taken place and each has had its proceedings published. The first was simply entitled *Comparative virology* and was published in 1971, two years after the conference. It provides one of the first presentations of viruses grouped according to morphological and biochemical characteristics rather than by host and disease. Chapters on classification, bacteriophages, insect viruses, plant viruses, animal and human viruses are all treated comparatively.

Viruses, evolution and cancer was the title of the published proceedings of the 2nd International Conference (1974), which provides information on comparative viral oncology in human, animal and plant cells, comparative immunology of oncogenic viruses and viruses and evolution.

The theme of the 3rd International Conference, held in 1977, was *Viruses and environment*. The proceedings were published two years later (1979) and presented the current trends in virus research and discussed the implications and role of viruses in biological and environmental studies. Also discussed were the use of live viruses as vaccines, biological control agents and in genetic engineering.

All conferences so far held have taken place at Mont Gabriel, Quebec. The series is edited by E. Kurstak and K. Maramorosch and published by Academic Press. Each volume has a subject and author index.

The organizing committee of these conferences, the ICVO, have also been responsible for holding the Annual Munich Symposia in Microbiology. Five have so far taken place with the last being entitled *Comparative aspects of leukaemias, lymphomas and papillomas*. The proceedings were published commercially by Taylor & Francis (1980).

International Congress for Virology

The first International Congress for Virology was held in 1968. Its proceedings, entitled *International virology I*, were published by S. Karger in 1969 and edited by J. L. Melnick for the virology section of the, then, International Association of Microbiological Societies. The second and third congresses were held in

Budapest in 1971 and Madrid in 1975 respectively. Their proceedings were published as *International virology II* and *III*. The Hague was the setting for the fourth congress in 1978, *International virology IV* being the published proceedings and the Centre for Agricultural Publishing and Documentation being the issuing body. Inside are found synopses of invited symposium papers and abstracts of workshop papers and posters. The symposium subjects were 'The taming of the viruses', 'Molecular mechanisms of viral oncogenesis', 'The ecology of viruses', 'Structure and assembly of viral particles', and 'The organization of viral genomes in relation to their replication and expression'. The 51 papers and posters in the workshop section were on a multitude of different subjects. Contributors are listed.

Society for General Microbiology Symposia

Surprisingly few of the SGM annual symposia have been devoted to virology, although the possibility of holding an annual one is under review. A number of the SGM's specialist groups such as Viruses; Genetics; Clinical Virology; Teaching; and the Irish and Scottish branches discuss virus research at ordinary and other meetings. Abstracts of these appear in the *SGM Quarterly*. Of those annual symposia which have been published the following are the most important:

Vol. 2	The Nature of Virus Multiplication (1952)
Vol. 9	Virus Growth and Variation (1959)
Vol. 18	Molecular Biology of Virus (1968)
Vol. 25	Control Processes in Virus Multiplication (1975)

4 The literature of virology

4.1	**Journals**	72
	The scientific journal	72
	Virology journals: some characteristics	72
	Individual virology journals	76
4.2	**Reviews**	87
	The information problem solved?	87
	Individual review and monograph series	88
4.3	**Books**	94
	Scientific books as information sources	94
	Individual virology books	97
	General	97
	History of virology	98
	Bacteriophages	99
	Chemotherapy (including interferon)	99
	Immunology	100
	Invertebrate virology	100
	Medical virology	100
	Methodology and diagnosis	101
	Molecular virology	102
	Plant virology (including fungal viruses)	102
	Slow viruses and viroids	103
	Specific groups	103
	Structure and morphology	104
	Taxonomy and classification	105
	Tumour virology	105
	Veterinary virology	106
4.4	**Conference proceedings**	107
4.5	**Dissertations**	108
4.6	**Reference works**	108
	Encyclopedias and handbooks	108
	Subject dictionaries	110
	Biographical dictionaries	112

4.1 JOURNALS

The term *journal* (or periodical) is assigned to publications that appear regularly, and usually, more frequently than annually. Newsletters, weeklies and bulletins are all forms of periodical. Unlike books (and this is their strength), periodicals provide a continuing stream of information, which enables them to react and report quickly on current events, issues or research. The article or scientific paper — the mainstay of the journal — owes its success to today's need for information to be encapsulated in a concise, and quickly and easily digested form. However, whilst the article is most prominent, journals do carry a wealth of different types of information. Indeed some journals, such as the *SGM Quarterly*, carry such a wide variety of services that they meet many people's entire information needs, providing research reports, calendars of meetings, job advertisements, notification of new books and periodical articles published, news of personalities and new products, state-of-the-art reviews, to mention but a few. They also provide a forum for debate and exchange of ideas via their editorial and correspondence columns. Exchanges, though generally of a more heated kind, do sometimes feature in the book review columns too. And all this between just one set of covers.

The Scientific Journal

There are some interesting facts about the scientific journal, a knowledge of which helps explain the functioning and use of virology journals.

The work on a paper published today was probably started 36 months ago; about 26 months ago the author started talking about his or her work at informal seminars; about the same time he started writing up his work for publication; approximately 12–15 months ago the author presented his paper at a conference; and shortly after this (8–12 months ago) the manuscript was submitted to the journal of his first choice (in about half the cases the author had already distributed copies of his manuscript to his colleagues).

On average one paper in three is rejected by the journal of the author's first choice. Rejection rates in biology (29%) are however appreciably better than those of mathematics (50%) but worse than for physics (24%).

The single author is a dying breed in science; most articles have two or three authors, and the trend is on the increase. Another trend is towards shorter papers, and this trend is most marked in the life sciences where the average paper is now only 6 pages in length. This is partly to do with the fragmented reporting of much research (i.e. work that would previously have been published in full in say an article of 12 pages in length is now reported in two parts). The practice of multiple publication (i.e. the same or a similar paper published in a number of different journals) adds to this paper inflation.

The average article will be read by about 100 readers and half the papers contained within a journal will be read in detail by no more than 1% of its readers — no paper will be read by more than 7%. The core of any science will be found in the articles of a very few journals and the majority of journals represent, in effect, the minority of the literature. The figure usually quoted is that 6% of the journals contain 90% of the valuable articles.

Virology Journals: Some Characteristics

There are just 13 scholarly journals that devote themselves entirely to the

Table 4.1a   Journals devoted entirely to virology (13)

Acta Virologica
Antiviral Research
Archives of Virology
Intervirology
Journal of General Virology
Journal of Medical Virology
Journal of Virology
Journal of Virological Methods
Revue Roumaine de Médecine, Série Virologie
Virology
Virus
Virusy i Virusnye Zabolevaniya
Voprosy Virusologii

dissemination of virological information (*Table 4.1*). However a much higher figure (5000) than this may actually carry virological information, though not necessarily regularly. The reason why so much virological information is carried in the journals of other related disciplines, is twofold. First, because of the subject's relative youth, virological information is still carried by the journals of its parental disciplines, chiefly biology and medicine. Secondly, there is a widespread interest in viruses throughout the life sciences and as a result reports on virus research can appear, albeit irregularly, in the journals of many of the subjects that make up the life sciences. If evidence of its widespread appeal is necessary one has only to look at the extent of coverage virology obtains in the prestigious science-wide journals such as *Nature* (*see Table 4.2* for details).

Whilst regular scanning of the relatively small population of specialist virology journals may be sufficient to keep students, teachers and practitioners abreast of the major developments in the field, the mainstream research workers will inevitably have to widen their reading to embrace journals in biochemistry, immunology, microbiology, medical and veterinary science. Fortunately, because a regular search involving the number of journals concerned — possibly 3000 — would be prohibitively expensive in terms of time (if indeed possible at all), papers of particular value are not scattered randomly throughout the journal literature. Invariably (and virology would seem to be no exception) the bulk of valuable items are concentrated in a relatively small and, importantly, manageable number of journals. This phenomenon largely obtains because there is so much duplication and redundancy in the literature; there are few journals that provide truly new, authoritative or valuable information, many contenting themselves with reporting what others have already said much better or just publishing very run-of-the-mill papers. Much publishing of course goes on for publishing's sake, communication being something of a secondary consideration.

Table 4.1b Journals of related disciplines regularly carrying many articles on virology (25)

Avian Diseases
Avian Pathology
Australian Veterinary Journal
Comparative Immunology, Microbiology and Infectious Diseases
Cornell Veterinarian
FEBS Letters
Infection and Immunity
Journal of Biological Standardization
Journal of Clinical Microbiology
Journal of Comparative Pathology
Journal of Immunology
Journal of Infectious Diseases
Journal of Invertebrate Pathology
Journal of the National Cancer Institute
Journal of Wildlife Diseases
Microbios
National Institute of Animal Health Quarterly
Netherlands Journal of Plant Pathology
Phytopathology
Plant Disease
Proceedings of the National Academy of Sciences
Proceedings of the Society for Experimental Biology and Medicine
Veterinary Microbiology
Veterinary Record
Zentralblatt für Bakteriologie, Mikrobiologie und Hygiene

One way of establishing what are the 'core' journals of a subject is to examine the field's abstracting service and determine which journals obtain best coverage — the assumption being of course that the more times the abstracting service sees fit to abstract a journal then the greater the importance of that journal's papers. The argument is flawed in some ways (i.e. a journal with more articles in it is statistically more likely to be ranked more highly; the method also takes no account of an abstracting service's particular historical preoccupations) but nevertheless has much truth in it.

Taking the 1979 volume of *Virology Abstracts* as our source (7000 abstracts and 5000 journals represented) and ranking the journals according to frequency of

Table 4.1c Journals of related disciplines regularly carrying a few but important articles on virology (19)

Annals of Neurology
Antimicrobial Agents and Chemotherapy
Biochimica et Biophysica Acta. Nucleic Acids and Protein Synthesis
Biochemical and Biophysical Research Communications
Cancer Research
Cell
International Journal of Cancer
Journal of Biological Chemistry
Journal of Experimental Medicine
Journal of General Microbiology
Journal of Molecular Biology
Journal of Neurological Science
Journal of Pediatrics
Lancet
Molecular and General Genetics
Nature
New England Journal of Medicine
Nucleic Acid Research
Science

coverage, we obtain the results shown in *Table 4.2*. The bunching effect can quite easily be seen from the above table, with just 3 journals (fewer than 1% of the journals abstracted) accounting for over 15% of the abstracts and just 10 journals accounting for one-quarter. Clearly *Virology* and *Journal of Virology* are virology's leading journals.

A more qualitative approach to calculating the value of a journal can be made by ranking journals according to the number of references made to them in the review articles of *Advances in Virus Research*, where one assumes that only the seminal articles published that year (it being an annual) are cited. The results of such an exercise based upon the 1979 volume of *Advances in Virus Research* are given in *Table 4.3*. What is immediately apparent from these results is that when it comes to qualitative considerations the bunching effect is even more marked, with just 2 journals accounting for a quarter of all the citations and 6 journals accounting for half. Interestingly this survey confirms the supremacy of *Virology* and *Journal of Virology*. In fact the consensus is even greater than that, with 5 out of the 6 top-placed titles being common to both lists.

Whilst the above two exercises are useful in that they help identify what are *generally* regarded to be the most important journals in the field, it does

Table 4.1d Journals of related disciplines occasionally carrying articles on virology (20)*

American Journal of Epidemiology
American Journal of Medical Science
American Journal of Public Health
American Journal of Tropical Medicine
American Journal of Veterinary Research
Annales de l'Institut Pasteur
Bulletin of the WHO
Canadian Journal of Microbiology
European Journal of Biochemistry
Federation of American Societies for Experimental Biology. Federation Proceedings
FEMS Letters
Gene
Immunology
Journal of the American Medical Association
Journal of the American Veterinary Medical Association
Journal of Tropical Medicine and Hygiene
Medical Microbiology and Immunology
Phytopathologische Zeitschrift
Postgraduate Medical Journal

*Bibliographic details not supplied unless the journal is mentioned in the text.

discriminate against the specialist journal that might not be so widely used or read but nevertheless contains information not usually found elsewhere. Neither do the analyses point to the relative subject strengths of even the more general journals listed in the tables. To compensate for this, *Table 4.4* lists those journals which contribute most significantly to the various specialisms found within virology. The table is based with some modifications on data obtained from *Virology Abstracts* (1980).

Individual Virology Journals (see bibliography entries 32–92)

The journal title abbreviations follow the usage of *Index Medicus*.

Acta Virologica (*Acta Virol.*)

A leading and well-established Czech journal containing approximately 15 original virus papers, book reviews and announcements in each issue. Subject material is of an immunological and sometimes chemotherapeutic nature and contributed mainly by Czech and Russian authors; occasionally however American, British and Japanese authors make their appearance. Retrospective searching is made easier by the provision of an annual index. Two language editions of the journal are published, one in Russian and one in English.

Table 4.2 Journals most frequently abstracted by Virology Abstracts

Rank	Journal title	Abstracts listed in Virology Abstracts (%)	Cumulative (%)
1	Virology	6.8	6.8
2	Journal of Virology	4.9	11.7
3	Journal of General Virology	3.5	15.2
4	Proceedings of the National Academy of Sciences	1.9	17.1
5	Lancet	1.6	18.7
6	Nature	1.6	20.3
7	Voprosy Virologii	1.5	21.8
8	Infection and Immunity	1.3	23.1
9	Molecular and General Genetics	1.1	24.2
=10	Journal of Biological Chemistry	1.0	25.2
=10	Intervirology	1.0	26.2
=10	Cell	1.0	27.2

Antimicrobial Agents and Chemotherapy (*Antimicrob. Agents Chemother.*)

This journal is devoted exclusively to all aspects of antimicrobial agents, anticancer agents and chemotherapy. Published by the ASM, each issue contains around 30 papers of interest to research workers in the pharmaceutical industry, microbiologists and specialists in infectious diseases. The papers on the chemotherapy of viral diseases are a little thin on the ground but are generally of much significance.

Antiviral Research (*Antiviral Res.*)

Youngest journal devoted to virology, it contains about 7 original and review articles pertaining to the control of virus infections in man, animals and plants by vaccines, serotherapy, physiological and antiviral mechanisms of host and antiviral chemotherapy. The first issue included articles on natural killer cells in relation to virus infections, interferon, chemotherapy of herpesvirus, bovine leukaemia infection, hepatitis and polio vaccination.

Archives of Virology (*Arch. Virol.*)

The first journal entirely devoted to virology, and called *Archiv für die Gesamte Virusforschung* as recently as 1975. The new name provides a better pointer to its contents, which are entirely in English. It usually contains some 10 original

Table 4.3 Journals most frequently cited in Advances in Virus Research

Rank	Journal title	Citations made in Advances in Virus Research (%)	Cumulative (%)
1	Virology	15.0	15.0
2	Journal of Virology	10.6	25.6
3	Proceedings of the National Academy of Sciences	7.8	33.4
4	Journal of General Virology	7.7	41.1
5	Nature	4.4	45.5
6	Journal of the National Cancer Institute	4.3	49.8
7	Journal of Experimental Medicine	3.9	53.7
8	Journal of Immunology	3.7	57.4
9	Archives of Virology	2.2	59.6
10	Journal of Molecular Biology	2.1	61.7

communications including brief reports in the areas of viral replication, tissue culture, immunology and characterization. The papers are concise and of a fairly general nature. A useful feature is the announcement of titles of articles to be published in future issues. Main contributors are the USA, the UK, Japan and France. A testament to the journal's importance is the fact that at least eight abstracting/indexing services see fit to include its articles. It must be regarded as expensive by the standards of other virology journals. An index is published.

Biochimica et Biophysica Acta (*Biochim. Biophys. Acta*)
This well-established Dutch journal appears under seven guises of which the one subtitled 'Nucleic acids and protein synthesis' contains the most virology, amounting to 1 or 2 important virus papers out of the 25 it publishes per issue. The virus research covers physicochemical studies, replication, oncology and phage genetics. It has a cumulative index and with a fortnightly publishing cycle is one of the most frequently appearing virology journals.

Cancer Research (*Cancer Res.*)
Official organ of the American Association for Cancer Research and co-sponsored by the US government, the Japanese Foundation for Cancer Research and Kureha Chemical Industry Co. Ltd. It is a large journal containing anything from 50 to 100 papers per issue with a small but significant virology content, concerned largely with the biochemical and immunological properties of tumour viruses. It is *Virology Abstracts*' most frequently covered journal in the field of

Table 4.4   Journals specializing in the various sub-fields of virology*

Subject	Journals
Virus taxonomy and classification	*Phytopathology* *Intervirology* *Journal of General Virology*
Methodology and tissue culture studies	*Journal of General Virology* *Virology* *Infection and Immunity* *Voprosy Virologii* *Acta Virologica*
Physicochemical properties, structure and morphology	*Virology* *Journal of Virology* *Journal of General Virology* *Proceedings of the National Academy of Sciences*
Replication cycle	*Virology* *Journal of Virology* *Journal of General Virology* *Nature* *Molecular and General Genetics*
Virus genetics including virus reactivation	*Virology* *Journal of Virology* *Molecular and General Genetics* *Journal of General Virology* *Gene*
Immunology	*Infection and Immunity* *Journal of General Virology* *Journal of Immunology* *Journal of Medical Virology* *Lancet* *Voprosy Virologii*
Phage–host interactions	*Journal of Bacteriology* *Molecular and General Genetics* *Journal of General Microbiology*
Antiviral agents	*Antimicrobial Agents and Chemotherapy*
Tumour virology	*Cancer Research* *International Journal of Cancer* *Journal of the National Cancer Institute* *Journal of Virology* *Virology* *Proceedings of the National Academy of Sciences*
Viral infections of man	*Lancet* *Journal of Pediatrics* *New England Journal of Medicine*

Table 4.4 (continued)

Subject	Journals
Viral infections of man (*continued*)	*American Journal of Public Health* *Medical Microbiology and Immunity* *Journal of Medical Virology*
Diseases associated with slow viruses	*Neurology* *Annals of Neurology* *Journal of Neurology*
Viral infections of animals	*Veterinary Record* *Australian Veterinary Journal* *Cornell Veterinarian* *Research in Veterinary Science* *National Institute of Animal Health Quarterly* *Avian Diseases*
Viral infections of invertebrates	*Journal of Invertebrate Pathology*
Viral infections of plants	*Plant Disease Reporter* *Phytopathology* *Netherlands Journal of Plant Pathology* *Acta Pathologica*

*Generally speaking those listed first are the most productive.

oncology. It also includes correspondence, reviews, meeting reports and announcements. It has a subject and author index.

Cell (*Cell*)

Frequently cited by virology authors, this journal contains reviews, mini-reviews, up to 30 original research articles, book reviews and announcements. Each issue contains about 5 papers on aspects of viral replication and genetics, with tumour virus models used extensively. Articles accepted for publication allegedly appear within three months of submission.

Comparative Immunology, Microbiology and Infectious Diseases (*Comp. Immunol. Microbiol. Infect. Dis.*)

Fairly new international journal with a relatively small circulation aimed at both medical and veterinary researchers and practitioners. Contains about 15 papers per issue of which half may be of interest to virologists working in the area of applied immunology and comparative pathology. It features special issues devoted to specific topics such as 'Herpesviruses' and 'Animal and human influenza'.

Infection and Immunity (*Infect. Immun.*)

Important American journal with a large audience containing 45–60 papers per issue, of which a quarter relate to virology. These articles can be located under one of five subdivisions entitled 'Viral infection and immunity'. In addition to original research papers, notes and announcements are also furnished. As the title indicates, the journal publishes mainly in the area of immunology; however articles do appear on methodology and on virus diseases of animals. It has an author and subject index in each volume and also a contents list of issues comprising a volume.

International Journal of Cancer (*Int. J. Cancer*)

The 18 original research papers (in English or French) each issue carries are, usually, equally divided between 'Human' and 'Experimental' cancer. Two-thirds of these papers may be of virological interest, concerned chiefly with viral oncology and immunology. The herpes simplex virus has received extensive coverage of late. The journal possesses a semi-annual index and is also available in microform. Prior notification of the contents of the next issue is usefully provided. The journal takes its current-awareness function seriously, publishing the contents lists of other cancer journals.

Intervirology (*Intervirology*)

An avowedly international journal of the virology section of the International Union of Microbiological Societies. It contains approximately 14 original reports per issue, of which one may be a review article, usually written by the relevant study group of the International Committee on Taxonomy of Viruses (ICTV). An entire issue may be devoted to a report of the work of the ICTV summarizing the state of approved virus nomenclature and taxonomy. The journal covers most areas of animal virology though virus structure and taxonomy, replication and plant virology get special attention. Contributors are mainly North American, the remainder being evenly spread throughout the world. As well as the research reviews, announcements, book reviews and bibliographies can be found. Available in microform.

Journal of Bacteriology (*J. Bacteriol.*)

This monthly journal of the ASM contains between 50 and 70 original papers of which a small number are usually concerned with bacteriophages and found under the subheading of 'Genetics and molecular biology'. The journal has a relatively large circulation of around 10 000 and is indexed extensively by bibliographic periodicals. Each issue carries an author index which is cumulated at the end of each volume, where a subject index will also be found.

Journal of Biological Chemistry (*J. Biol. Chem.*)

This large, academic journal of the American Society of Biological Chemists contains 70–80 original papers per issue, of which an average of 5 in the area of animal and bacterial virus replication and biochemistry will be of interest. These will be found under the subject category 'Nucleic acids, protein synthesis and molecular genetics'. This journal was the most cited scientific periodical of 1979, a feat no doubt helped by its interdisciplinarity and size. It has a cumulative author index and is available in microform.

Journal of Clinical Microbiology (*J. Clin. Microbiol.*)

Partly as a result of its appeal to the large number of practitioners in the field, this journal has a large circulation by microbiological standards. It concentrates on the applied microbiological aspects of human and animal infections, particularly in relation to their aetiology, diagnosis and epidemiology. As many as a quarter of the 40 original research articles published in each issue will be of some relevance to virology. There is a bias towards immuno methodological aspects, with descriptions of serological tests such as optical immunoassays and electron microscopy. Virology will be located under headings 'Virology, rickettsia, chlamidiae', 'Epidemiology', 'Immunology', 'General clinical microbiology' and 'Clinical veterinary microbiology'. Other contents include announcements and semi-annual indexes. It is available in microform.

Journal of Experimental Medicine (*J. Exp. Med.*)

A long-established journal which has figured strongly in virology's development and is thus frequently cited for historical reasons. It still publishes about 5 original virus research papers per issue, mainly in the area of immunology and particularly vaccination therapy. Can be obtained in microform.

Journal of General Microbiology (*J. Gen. Microbiol.*)

The journal of both the Society for General Microbiology (SGM) and the Federation of European Microbiological Societies (FEMS). It contains about 30 original articles per issue, 3 of which are usually in the area of phage genetics and viral immunology. It carries a cumulative index.

Journal of General Virology (*J. Gen. Virol.*)

Like its sister journal above, it is an organ of both the SGM and FEMS. It publishes accounts of original virus research mainly in the field of animal virology, although plant viruses are also relatively well catered for. Methodological articles are a feature of the journal and additionally 1–2 papers an issue concern bacteriophages. In all, approximately 24 papers of both an applied and pure nature are published per issue of which 1 or 2 may be review articles from invited experts. Its importance in the field can be gauged by the fact that it is the journal that receives the third largest number of abstracts from *Virology Abstracts*. In 1979 contributions came from 31 countries although the UK, USA and Japan dominated. Articles are high-powered and fairly concise.

Journal of Immunology (*J. Immunol.*)

This long-established, widely circulated journal is an organ of the American Association of Immunologists. It is a large journal that contains as many as 70–80 original papers per issue covering all aspects of immunology. Virus-orientated research accounts for about 10% of most issues and can be located under the subject category 'Viral and microbial immunology'. Tumour viruses figure prominently in the reported research. It is available in microform.

Journal of Invertebrate Pathology (*J. Invertebr. Pathol.*)

Published under the auspices of the Society for Invertebrate Pathology, this journal provides coverage of a neglected but emerging branch of virology, namely viruses of invertebrates. Of the 15 or so papers published per issue 2–3 will be on the replication and infectivity of such viruses as single embedded polyhedrosis virus, usually in insect cell lines. The articles are fairly general and quite readable by the layman. In addition there are book reviews and a cumulative author and subject index in every volume.

Journal of Medical Virology (*J. Med. Virol.*)
A fairly new journal containing only about 8–10 papers an issue and specializing in viral infections of man, with hepatitis obtaining a very detailed coverage. The emphasis is largely on applied research, mainly in the field of immunology on topics such as humoral and cell-mediated response. There is some coverage of treatment and prevention. Authors are mainly European. On the inside of the back cover are printed the titles of papers due to appear in future issues.

Journal of Molecular Biology (*J. Mol. Biol.*)
A frequently appearing prestigious British journal containing 16–20 papers per issue in the ubiquitous area of molecular biology. Papers on animal virus and phage replication constitute the viral contribution of which 1–2 appear in each issue. Also covers announcements of meetings and has a cumulative index.

Journal of the National Cancer Institute (*J. Natl Cancer Inst.*)
The organ of the National Cancer Institute and largely devoted to reporting the work of this body's employees. The 25 or so original communications each issue carries are divided into 'Investigations into man' and 'Investigations into non-human systems', usually in favour of the latter. The viral interest in the journal rests in about 10% of its contributions, taking the form of tumour virus immunology and genetics. The journal also features reviews, symposia announcements and news items.

Journal of Virological Methods (*J. Virol. Methods*)
A very recent addition to virology (1980), the journal so far contains an average of 6 original papers an issue and is mainly of medical interest. Immunological methods are featured strongly; biochemical ones obtain coverage but on a somewhat reduced scale. Accounts are applied and very readable. Its publication demonstrates how virology — once a specialism within microbiology — is now developing its own specialisms.

Journal of Virology (*J. Virol.*)
With *Virology*, undoubtedly one of the leading journals in the field. Citation behaviour and frequency of coverage by *Virology Abstracts* both confirm this. It has a relatively large readership for a specialist journal (about 6000). It contains around 40 mainstream virology papers an issue, covering such fundamental aspects of virology as physicochemical and morphological studies, replication, genetics and tumour virus research. Animal virus research predominates, some phage but little in the way of plant virus research. This ASM journal is fairly academic and aimed at the initiated. Over 75% of its contributors are from the USA, which makes it the virology journal in which Americans have the greatest monopoly. The UK, France and the Netherlands make up the rest. It is available in microform and carries a semi-annual index to aid literature searching.

Journal of Wildlife Diseases (*J. Wild. Dis.*)
Fairly well-established journal of the Wildlife Disease Association (WDA) covering, amongst other microorganisms, viruses of a relatively exotic nature: those of fish, reptiles, wild birds and mammals. Each issue carries about 3–4

original papers in virology, which may be found under the appropriate subheading. Also contains minutes of the meetings of the WDA and is indexed.

Lancet (*Lancet*)

One of the oldest scientific journals (and 8th most cited in 1979) in existence. Unusually, it can be obtained in two editions, UK and North American. It differs from the majority of scientific periodicals in that it has adopted a journalistic-type approach to reporting research within its magazine format. News, editorials, announcements, surveys, book reviews, topical review articles are presented. Most issues have 4–5 original articles, and some 30 brief communications in the form of correspondence which occasionally features virological topics. However, what the issues lack in number they certainly make up in quality. The material covered is, without exception, of medical importance and is indexed semi-annually.

Microbios (*Microbios*)

An international biomedical research journal devoted to fundamental studies of viruses and other microorganisms. Work on cell–virus interactions is frequently reported, as are biochemical and biophysical studies. The journal's speciality is the rapid publication of papers. Its sister publication *Microbios Letters* publishes articles even more quickly, though at a cost — that of design and attractiveness — and is intended to augment *Microbios*.

Molecular and General Genetics (*Mol. Gen. Genet.*)

This was the first journal devoted to genetics and was previously known as *Zeitschrift für Vererbungslehre*. Approximately 16 of the 90 or so original articles describe research in the field of bacteriophage genetics. It is in fact the main source of information on bacteriophage research.

Nature (*Nature*)

The most prestigious and third most cited scientific journal in 1979. So prestigious and wide in appeal is this general scientific journal that it is listed within the top ten most cited journals of most scientific disciplines. Research workers in all branches of science hope ultimately to publish here for a number of motives amongst which are status and rapidity of publication. Papers must be of great importance or of extreme novelty. The majority of major discoveries in virology were first published in *Nature*. It is of similar style to *Lancet*, with original articles and letters, news editorials, book reviews and announcements. Again the journalistic style and format helps assimilation; so too does the brevity of the articles. Possibly science's most abstracted journal.

Neurology (*Neurology*)

This is the official monthly journal of the American Academy of Neurology. In addition to the original papers, of which there are some 12 per issue, there are a similar number of 'features'. The virology content, amounting to 1–2 papers, is concerned with viruses of the nervous system. The journal also contains correspondence, opinions and a calendar of meetings.

New England Journal of Medicine (*N. Engl. J. Med.*)

American equivalent of *Lancet* but older and with a larger circulation. Contains original articles, correspondence, book reviews, news and announcements. The virology articles published are of course primarily of medical importance and cover isolation, identification and epidemiology of viruses of man. Contents are indexed semi-annually.

Phytopathology (*Phytopathology*)

The USA's leading research journal in the field of plant pathology. Four out of the 24 original papers carried in each issue concern viral infections of plants and, occasionally, fungi. Relevant articles can be found scattered under a number of headings: 'Etiology', 'Genetics', 'Cytology and histology', 'Ecology and epidemiology', 'Techniques', 'Resistance', 'Disease detection and loss', 'Disease control' and 'Pest management and vector relations'. The journal specializes in the characterization of plant viruses. Other contents include book listings, news, reviews and notes. It has indexes and cumulative indexes.

Plant Disease (*Plant Dis.*)

Formerly *Plant Disease Reporter* and published by the US Department of Agriculture, it is now the official journal of the American Phytopathological Society and sister journal of *Phytopathology*. It differs from the latter in that it adopts a magazine format. It has editorials, letters, news, features, announcements. About a quarter of the 15 original research papers it publishes every issue are of interest to plant virologists in the areas of genetics, infection studies and plant virus taxonomy. It also features new diseases and epidemics. It is subtitled *The International Journal of Applied Plant Pathology*.

Proceedings of the National Academy of Sciences of the United States of America (*Proc. Natl Acad. Sci. USA*)

Rivals *Nature* as the most prestigious scientific journal and is just one place behind it in the 1979 *Science Citation Index* rankings. Despite its wide coverage the journal can be considered to be one of the four most important sources of virological articles. It comes in two editions: *Physical Sciences* and *Biological Sciences*. Only the latter is of concern to us and is split into a number of subject headings, of which 'Biochemistry', 'Cell biology', 'Immunology', 'Medical sciences' and 'Microbiology' may yield papers of great interest to virologists, but most such papers will be found under 'Viral genetics' and 'Oncology'. Usually 10 out of the 100 original articles published in each issue are of relevance. Like *Nature* it publishes only papers of exceptional importance or novelty. Unlike *Nature*, however, prospective authors must be associates of the Academy, or publishing on behalf of one, or have received an invitation from the editors. Papers prepared at Academy symposia, special lectures or papers on subjects of special interest are also published. Publication is very rapid and may be within 10–15 weeks of submission.

Proceedings of the Society for Experimental Biology and Medicine (*Proc. Soc. Exp. Biol. Med.*)

Published for the Society for Experimental Biology and Medicine, it carries 10–20 original research papers in the field of biomedicine. The virology content is restricted to 2–3 papers an issue, conveniently located under the subject heading 'Virology'. Articles placed under the heading 'Immunology' should also be scanned as they may concern human viral vaccines and associated immunology.

Revue Roumaine de Médecine—Virologie

Translates as the *Romanian Journal of Medicine—Virology*, but has also been known as *Revue Roumaine de Médecine, Série de Virologie; Revue Roumaine de Virologie*; and *Revue Roumaine d'Inframicrobiologie*. This quarterly journal, published by the Academie des Sciences Médicales, is, surprisingly, mainly in English with the odd paper being in French, German or Russian. The journal has a medical bias, with papers, of which there are about 6 per issue, on applied immunological aspects of hepatitis, herpes, myxoviruses, paramyxoviruses and others. The occasional paper may be on animal viral diseases such as equine infectious anaemia. A couple of reviews usually accompany the papers and there are also book reviews, scientific events, correspondence and a listing of doctoral theses.

Veterinary Record (*Vet. Rec.*)

This journal is the major source of information on the viral infections of animals. It uses a mixed format of news and reports, editorials, book reviews, articles, short communications and letters to disseminate this information. It carries information on topical virus diseases of domestic animals and livestock from the point of view of new isolations, outbreaks and treatment. *Veterinary Record* is supplemented by another BVA publication, *Research in Veterinary Science*, which is a more orthodox scientific periodical containing original research in animal health with particular emphasis on microbiology.

Virology (*Virology*)

Virology's premier journal. Twelve years older than the *Journal of Virology*, it started publishing during the golden age of virology. It contains up to 30 high-quality virological communications per issue. It specializes in many aspects such as fundamental biochemical and structural investigations, virus replication and genetics as well as cell–virus interactions. Whilst nearly 75% of the material published is on the subject of animal viruses, plant and bacterial viruses get a better coverage than in most other virological journals. Similarly the journal is not so US-dominated; European, Australian, South African and Japanese authors are frequent contributors. The importance of the journal is reflected in the number of abstracting and indexing services (12) that regularly scan its content. It contains an author and subject index at the end of each volume. Like the *Journal of Virology* it publishes very much at the postgraduate level.

Virus (*Virus*)

Little-known but long-established journal of the Society of Japanese Virologists. Its value as an information source is somewhat circumscribed by the language of its contents — Japanese (English summaries are however provided). The 6 or so articles — comprising 2 reviews, 2 original papers and some miscellaneous material in each issue — are of a methodological type, mainly in the area of chemotherapy and serology. Contributors are predominantly Japanese. It is published with an author and subject index.

Virusy i Virusnye Zabolevaniya

This little-known and poorly printed journal of the Odessa Institute for Virology and Epidemiology, translating as 'Viruses and Virus Diseases', is entirely in Russian. It replaced an earlier publication known as *Ostrye Respiratornye*

Zabolevaniya in 1972. The journal contains about 20 short papers on aspects of medical virology and a small abstracting service covering some 50 articles per issue.

Voprosy Virusologii (*Vopr. Virusol.*)

A cover-to-cover English translation began in 1980 entitled *Soviet Progress in Virology*. This resumed publication of an English translation previously known as *Problems of Virology*, printed between 1956 and 1961. Each issue begins with one or more survey articles on topical subjects such as hepatitis, followed by about 25 original research articles reporting research on a broad front: biochemical and biophysical studies; replication; immunology; and infections of man, their diagnosis, treatment and prophylaxis. The journal also publishes discussions and critiques, book reviews, news and announcements. The interest in Russian research is demonstrated by the fact that about a dozen abstracting and indexing services see fit to cover its articles.

.2 REVIEWS

The Information Problem Solved?

Scientists do not generally suffer from a shortage of information; indeed, most are inundated with it. As a consequence they usually require help not so much in finding more information but in selecting the most valuable or relevant information from a literature that is said to be doubling in size every ten or fifteen years. (Norrby (1982) goes as far as to say the literature of virology is doubling every five years!) Abstracting or indexing services cannot provide the answer as they really only order the literature by constructing search pathways through it — they can do nothing in the way of reducing its bulk; indeed, because they have the unfortunate habit of paralleling growth in the primary literature, they actually add to it. Furthermore it has been clear for many years now that the scanning of indexes and abstracts is proving too much for many workers, and a more easily digestible alternative is required.

What has been seen increasingly as the solution to the information problem is the review. Reviews take the form of a critical summary of current developments in a particular field and are normally written by eminent specialists (the level of eminence, of course, very much depends on the prestige of the journal). The philosophy behind the review is a simple and worthy one: from a literature of say a thousand or so research papers the reviewing author selects perhaps a hundred of the most noteworthy, filtering out as a result the run-of-the-mill and less valuable. The prospect of someone — and an important someone at that — reducing the literature to such manageable proportions is a particularly attractive one to the besieged scientist, to whom the literature search is only a secondary, possibly tiresome, activity anyway.

The value of a review hinges upon the author's critical powers and, ultimately, the quality of his selection. To make a valid selection in the first place the author must have surveyed the total pool of potentially relevant papers. Now to do this properly requires not only plenty of time but also considerable bibliographic expertise (in first establishing that a relevant paper exists and then locating and obtaining it). In such an interdisciplinary field as virology, where items may emerge from any number of possible outlets, the task is that much harder.

It must be remembered however that exhaustiveness is not the chief criterion by which a review is judged; nothing is in fact duller and less helpful than those meticulous papers (and virology has its lion's share) that mention every fact and paper no matter how trivial — a mention of 700 or so papers is by no means uncommon. And when one considers that the only form of criticism employed by authors of reviews is omission then it becomes clear that many so-called reviews are nothing more than annotated bibliographies masquerading as reviews. Adding the fact that personal bias can intrude into the review, one can easily see that the review is not the panacea it was once thought to be.

Reviews may appear in a number of different locations: as articles in journals, chapters in books, papers in conference proceedings or grouped conveniently together in publications of their very own — the annual review of progress. However, whatever the location, their function, purpose and content are essentially the same. What might differ however is their topicality — the most up-to-date are likely to appear as papers given at conferences and the least current in textbooks, and complexity — journal reviews being most likely to be written at an advanced level.

Journals regularly publishing good review articles include: *Review of Plant Pathology* (6); *Comparative Immunology, Microbiology and Infectious Diseases* (5); *Intervirology* (4); *Journal of General Virology* (3); and *Nature* (10). The figures in brackets provide an indication of the number of reviews that are published each year.

One of the easiest ways of locating a review on a particular topic is to go to one of the serials that are entirely devoted to reviews. There are six that exclusively cover virology. In the main they are all written for the more general reader and appeal in particular to students, teachers and researchers working outside their own specializations. In all three cases reviews offer a convenient starting-point to a retrospective literature search.

The same heterogeneous mixture that forms the periodical literature of virology — indeed which lies at the very heart of the subject itself — is evident too in the review literature, with the fields of cancer research, immunology, genetics, microbiology, phytopathology, cytology, molecular biology and experimental pathology all contributing review journals of varying levels of relevance.

Reference

Norrby, E. (1982). Virology. In *Biology in profile*. Ed. by P. N. Campbell. Oxford: Pergamon.

Individual Review and Monograph Series *(see* bibliography entries 93–102)

Advances in Virus Research *(Adv. Virus Res.)*

Oldest (began 1953) and best known of the annual reviews in virology, it reached volume 26 in 1980. Lately, in an attempt to get to grips with a rapidly expanding literature and to maintain topicality, Academic Press has published two volumes a year. An average of 8 reviews an issue are published on a great diversity of virological topics. Many, in fact, turn out to be exotic and unpredictable. Plant and insect virologists will be grateful for the overdue attention they have received in recent editions. Animal virology is liberally catered for but those requiring information on human virus diseases are best served elsewhere.

The reviews generally are 50 pages in length, but occasionally run to 100 pages. Each review has an extensive bibliography, some having as many as 500 references. As usual American authors predominate with Russian, British and other Europeans somewhat less well represented. The series is edited by M. A. Lauffer, F. R. Bang, K. Maramorosch and K. Smith.

The great range of topics included can be best illustrated by listing the contents of a recent edition (1979):

> Guidelines for bacteriophage characterization/The tobraviruses/The origin of multicomponent small ribonucleoprotein viruses/Small isometric viruses of invertebrates/Reptilia-related viruses/Genetics of resistance of animals to viruses/Neutralization of animal viruses/Sendai viruses

Each edition contains a subject and author index.

CMI/AAB Descriptions of Plant Viruses

This series of looseleaf sheets is sponsored jointly by the Commonwealth Mycological Institute and the Association of Applied Biologists. It provides a standardized, authoritative description of viruses of plants. Each description consists of 2-3 pages of text plus 1 page of illustration and is prepared by an expert in the field. It includes information on disease caused; host range and effects on plants, modes of transmission; purification; serology, properties, composition and structure of particles.

The most recent of 13 sets published at the end of 1979 is:

Set 13
201 Lilac ring mottle
202 Lilac chlorotic leafspot
203 Andean potato mottle
204 Blueberry shoestring
205 Sonchus yellow net
206 Potato black ringspot
207 Beet yellow stunt
208 Satsuma dwarf mosaic
209 Cowpea severe
210 Beet curly top
211 Alfalfa latent
212 Cowpea mottle
213 Cucumber mosaic
214 Tymovirus group
215 Bromovirus group

Comprehensive virology

Whilst strictly not a review serial in the mould of *Advances in Virus Research* — it is more encyclopedic in content — *Comprehensive virology* is included here because it is of much the same purpose. This 18-volume, 6000-page compendium was completed in 1981. The editors, H. Fraenkel-Conrat and R. R. Wagner, deserve credit for their monumental achievement, which took less than eight years, thus minimizing any serious outdating of earlier issues.

Volume 1 contains an alphabetic catalogue and description of almost all viruses of bacteria, plants, invertebrates and vertebrates. Volumes 2–4 essentially deal with the reproduction of small and intermediate RNA viruses, DNA animal viruses and large RNA viruses. The general structural principles and assembly of virus particles are described in volumes 5 and 6, while in volume 7 the series of volumes on reproduction is concluded with a specific examination of DNA bacteriophages. Volume 8 covers the biological properties of DNA bacterial viruses from the point of view of regulation and genetics. A comprehensive analysis of the genetics of all animal viruses is provided in volume 9, and volume 10 deals with the regulation of viral gene expression. Volume 11 returns to the subjects of regulation and genetics, with specific reference to plant viruses. Volume 12 examines the viruses of algae, fungi and invertebrates. In volume 13 are contained various topics relating to the structure and assembly of viruses, including incidentally the complete sequencing of a viral RNA. In volume 14 special or newly characterized viruses are singled out for attention. Volume 15 looks at viruses and their role in immunology, and volume 16 reviews our knowledge of viral invasion persistence and certain diseases. With volume 17 presenting information on the biophysical and serological methods used in virus research, the series concludes with volume 18, dealing with cell responses to viral infection and other viral diseases.

The editors (Fraenkel-Conrat and Wagner) have further plans; rather than updating the series, they are preparing a group of books each dealing exhaustively with a specific virus family. The series will be simply entitled *The viruses* and will commence with three books on the Herpesviridae, to be edited by B. Roizmann.

Unlike most reviews, which select particular areas of interest (many being interdisciplinary) the *Comprehensive virology* series has undertaken to cover the entire field of virology. It has thus superseded, though rather expensively, some of the standard textbooks that have been around for some time.

Vol. 1 *Descriptive catalogue of viruses* (1974)
Vol. 2 *Reproduction: small and intermediate RNA viruses* (1974)
Vol. 3 *Reproduction: DNA animal viruses* (1974)
Vol. 4 *Reproduction: large RNA viruses* (1975)
Vol. 5 *Structure and assembly: virions, pseudovirions and intraviral nucleic acids* (1975)
Vol. 6 *Structure and assembly: assembly of small RNA viruses* (1976)
Vol. 7 *Reproduction: bacterial DNA viruses (1976)*
Vol. 8 *Regulation and genetics: bacterial DNA viruses* (1977)
Vol. 9 *Regulation and genetics: genetics of animal viruses* (1977)
Vol. 10 *Regulation and genetics: viral gene expression and integration* (1977)

Vol. 11   *Regulation and genetics: plant viruses* (1977)
Vol. 12   *Newly characterized protist and invertebrate viruses* (1978)
Vol. 13   *Structure and assembly: primary, secondary, tertiary, quaternary structures* (1979)
Vol. 14   *Newly characterized vertebrate viruses* (1979)
Vol. 15   *Virus–host interaction: immunity to viruses* (1979)
Vol. 16   *Virus–host interaction: viral invasion persistence and disease* (1980)
Vol. 17   *Methods used in the study of viruses* (1981)
Vol. 18   *Virus–host interaction: cell response to viral infection and disease* (1981)

Handbuch der Virusinfektionen Tieren

An encyclopedic work that manifests itself in the form of a series of books. It translates as the 'Handbook of viral infections of animals' and succeeds *Handbuch der Viruskrankheiten* by Gildemeister, Haagen and Waldmann, issued in 1939.

This colossal work, edited by V. H. Rohrer, comprises nine 900-page volumes and forms the most complete compilation of veterinary virology ever undertaken, although because it is in German, it is unlikely that many virologists are aware of it. In all there are 91 chapters on individual viral and rickettsial diseases of animals, including poultry, laboratory animals and fish, contributed by authors from France, Sweden, Finland, Bulgaria, Poland, South Africa, Egypt, Romania and Germany. Full lists of references are provided at the end of each chapter as well as indexes to authors cited in the text and a subject index.

Band   I   1967
Band   II   1967
Band   III 1968 (in two parts)
Band   IV   1969
Band   V   1969 (in two parts)
Band   VI   1978 (in two parts)

Methods in virology

This series is intended to provide a comprehensive treatise on methods used in the study of human, animal, plant, insect and bacterial viruses. It comprises six volumes of over 4000 pages compartmentalized into nearly 100 chapters, each by a different author or set of authors specializing in that particular field. Each has an author and subject index.

The first four volumes of this series were published in rapid succession (1967–68). Volume 5 appeared in 1971 with the twofold objective of updating the previous work and identifying techniques which would be useful in the future. Six years later the editors, K. Maramorosch and H. Koprowski, again felt the need to acquaint old and new readers with new and improved techniques, particularly those brought about by the widespread introduction of the electron microscope.

The themes of each volume are as follows:

Vol. 1   Basic methods of study such as tissue culture, virus transmission, maintenance of the host system (e.g. arthropods, plants and nematodes) and bacteriophage techniques (1967)

Vol. 2 Physicochemical methods of study such as ultracentrifugation, ultrafiltration and chromatography (1967)

Vol. 3 Quantitative and qualitative techniques for studying antigen and antibody, microscopic and electron microscopic methods (1967)

Vol. 4 Methods for the study of antiviral compounds and inactivation procedures, defective viruses, virus vaccine production and methods for storage (1968)

Vol. 5 Polyacrylamide gel electrophoresis, hybridization, immunoperoxidase techniques (1971)

Vol. 6 Immunofluorescence assay methods for viruses, viroids and viral components, immune electron microscopy and invertebrate cell culture for the study of both animal and plant viruses (1977)

Monographs in Virology

Probably the most varied and exotic of the virology monograph review series covering everything from bat viruses to viral taxonomy. J. L. Melnick is the overall editor of the series. The average length of monographs is about 100 pages, with a massive reference section accounting, sometimes, for half that length.

Previous monographs are:

Vol. 1 *Rhinoviruses* — D. Hamte (1968)

Vol. 2 *Enzyme induction by viruses* — S. Kit, D. R. Dubbs (1969)

Vol. 3 *Persistent and slow virus infections* — J. Hotchin (1971)

Vol. 4 *Viral structural components as immunogens of prophylactic value* — R. A. Neurath, B. A. Rubin (1971)

Vol. 5 *Classification and nomenclature of viruses* — P. Wildy (1971)

Vol. 6 *Moving frontiers in invertebrate virology* — T. W. Tinsley, K. A. Hassap (1972)

Vol. 7 *The agent of trachoma* — Y. Becker (1974)

Vol. 8 *Virus infections in bats* — S. E. Sulkin, R. Allen (1974)

Vol. 9 *Early interaction between animal viruses and cells* — K. Lonberg-Holm, L. Philipson (1974)

Vol. 10 *Cytopathology in viral diseases* — N. Cheville (1975)

Vol. 11 *Antiviral drugs* — Y. Becker (1976)

A cumulative author/subject index appeared in volume 9.

The series is aimed at the informed general reader.

Two further reports of the International Committee on Taxonomy of Viruses (*see* vol. 5), by F. Fenner (1976) and R. E. F. Matthews (1979) respectively, have been published by S. Karger.

Perspectives in Virology

This series of books contains the proceedings of the roughly biennial Gustav Stern Symposia, at which the current thinking of virologists engaged in the study of persistent virus infection, viral chemotherapy, immunology and other mainly medical topics is unveiled. Discussions following each review are verbatim reports and there are extensive references at the end of each discussion and illustrations liberally distributed.

Volume 10 (1978) departed from the earlier format of devoting the meeting to a central theme by considering recent advances in virology. The 14 articles, of about 15 pages each, included the first confirmed propagation *in vitro* of hepatitis B antigen, the genetic analysis of influenza viruses, an assessment of viral insecticides, new human virus diseases and human wart viruses.

Earlier volumes included:

- Vol. 5 *Virus-directed host response* (1967)
- Vol. 6 *Virus-induced immunopathology* (1969)
- Vol. 7 *From molecules to man* (1971)
- Vol. 8 *Persistent virus infections* (1973)
- Vol. 9 *Antiviral mechanisms* (1975)

The series is edited by M. Pollard and each volume contains an author/subject index.

Progress in Medical Virology (*Prog. Med. Virol.*)

This annual series, edited by J. L. Melnick, reached volume 25 in 1979. It deals predominantly with human viral diseases but has also carried articles on general virology and virus infections of non-human primates. Surveys of the current state of viral taxonomy are also included periodically.

The number of reviews per edition does not usually exceed 10, all with extensive references. Recent topics have included: 'Similarities and differences between viral and cellular membranes', 'Animal virus pseudotypes', 'Adeno-associated virus', 'Viral antibodies in multiple sclerosis', 'Enteroviruses in human diseases', 'Viral vaccines'.

'*Progress*' provides an up-to-date reference source for workers in the field but is also of value to those not directly involved. It is well illustrated and contains author and subject indexes.

Virology Monographs

This series continues *Handbuch der Virusforschung*, founded by R. Doerr. It is edited by C. Hallauer. In general each edition is devoted to one subject although occasionally this may be increased to two. While the area covered by the series is very varied, all the monographs so far have been concerned with mammalian viruses and in particular, human ones. The mean length of each review is about 150 pages and the reviews are followed by a very comprehensive bibliography of up to 500 or so references, and copious illustrations and tables.

The volumes so far are as follows:

- Vol. 1 *Echo viruses/Reoviruses* (1968)
- Vol. 2 *The simian viruses/Rhinoviruses* (1968)
- Vol. 3 *Cytomegaloviruses/Rinderpest virus/Lumpy skin disease virus* (1968)
- Vol. 4 *The influenza virus* (1968)
- Vol. 5 *Herpes simplex and pseudorabies viruses* (1969)
- Vol. 6 *Interferon* (1969)
- Vol. 7 *Polyoma virus/Rubella virus* (1969)
- Vol. 8 *Spontaneous and virus-induced transformation in cell culture* (1971)
- Vol. 9 *African swine fever virus/Blue tongue virus* (1971)

Vol. 10 *Lymphocytic choriomeningitis virus* (1971)
Vol. 11 *Canine distemper virus/Marburg virus* (1972)
Vol. 12 *Varicella virus* (1972)
Vol. 13 *Lactic dehydrogenase virus* (1975)
Vol. 14 *Molecular biology of adenovirus* (1975)
Vol. 15 *The parvoviruses* (1976)
Vol. 16 *Dengue viruses* (1977)
Vol. 17 *The nature and organization of retroviral genes in animal cells* (1980)

Others (*see* bibliography entries 103–116)

As well as the serials and monographs that have been included so far, which deal exclusively with virology, there are a number of others that regularly or occasionally include a review on some aspects of virology. The nature of the reviews are indicated by the title of the serial; thus 'The role of viruses in cancer', or 'Genetics of P_2 and related phages' will be found in those serials covering cancer research or genetics respectively.

The following can, generally, be expected to carry at least one or two reviews a year on some aspects of virology:

Advances in Cancer Research
Advances in Immunology
Advances in Veterinary Science and Comparative Medicine
Annual Review of Genetics
Annual Review of Microbiology
Annual Review of Phytopathology
CRC Critical Reviews in Microbiology
Current Topics in Microbiology and Immunology
Microbiological Reviews
Progress in Nucleic Acid Research and Molecular Biology
World Health Organization Technical Report Series

4.3 BOOKS

Scientific Books as Information Sources

A large number of quite different publications shelter under the umbrella term *book* (or monograph, to use a more technical description); the only common factor is their 'one-offness' (i.e. they are either completed in one volume or intended to be completed in a finite number of volumes).

The following publications may be considered to be books: pamphlets, encyclopedias, dictionaries, manuals, research reports, theses. Additionally conference proceedings, directories, government publications may or may not be regarded as books depending on their publication pattern (i.e. a proceedings published annually is a serial but one published just once is a book).

It is the intention to treat here only the 'simple' book; its other, more specialist manifestations are dealt with separately, elsewhere.

Attitudes to the book in the scientific community are indeed ambivalent. On the one hand the institution of the book, especially the hardback, commands general respect; and on the other books are widely suspect. This suspicion emanates from the fact that: (a) the book is not subject to the same degree of scrutiny as are journal articles; and (b) the latest information on a subject is often missing because a book's production and publication is a particularly lengthy and arduous procedure. It is for these reasons that books are often considered peripheral to the scientific literature. Yet it is our contention that the book is unnecessarily maligned and neglected, it being in fact a valuable and very accessible source of information.

Ambivalence towards the book is often illustrated by the most vocal critics, whose bookshelves contain a number of ancient, much loved and much used texts, the usage deriving from familiarity rather than a need for accuracy.

A good textbook differs from the most carefully chosen set of journal references in several obvious but nevertheless crucial ways. It will have a sense of history that cannot be conveyed in the confines of the journal article and it can also afford to be far more generous in its assessment of a topic than the latter. It can also be challenging, stimulating and ideally, original — a description, with originality aside, that can rarely be applied to a journal article. It is often *because* of the book's relative lack of scrutiny, mentioned above, that established views can be more easily challenged.

Virology, though still expanding, has an established body of information that, by and large, is accepted by its disciples and thus readily amenable to the transition into the more permanent form offered by the book. There are, of course, many more virology books than virology periodicals, though not nearly as many as there are virology articles. Well over 350 virology books have been published over the last two decades, with annual production accelerating to about 60 a year in recent years. (This total does not include the large number of books of other, related disciplines that have sections or chapters devoted to viral topics; the inclusion of these would increase the book population by at least fourfold.)

Whilst in many disciplines, textbooks are primarily educational tools for students, in virology they are largely used by research workers either to obtain an introduction to an unfamiliar subject area or, if sufficiently advanced, as a source of reference. This is mainly because virology is rarely studied in its own right at the undergraduate level; more normally it is taught as a component of a microbiology, medical or veterinary degree. It is unfortunate that the set books for the virology component of many of these courses are microbiological (i.e. virology will be found mixed in with bacteriology and mycology) as, in most cases, virology comes off the worse. The use of a microbiology textbook is often justified for economic reasons but in fact cheap and very good virology texts, many in paperback, do exist.

The majority of virology books are written by more than one author. Occasionally there can be as many as 20 contributors, usually under the control of one or two editors. Such practices can lead to confusion when attempting to trace these works because some libraries and information tools place them under the editor's name whilst others put them under the title or the first-named author.

The multi-author volumes occasionally suffer from a mixture of styles but do allow an extended and systematic exposition of the subject by experts in the field. Such books may be more precisely termed 'treatises' but will be considered in this section.

For reasons outlined earlier very few basic virology books are published that do not require a relatively sophisticated level of understanding; in particular, nearly all require as a prerequisite a fair knowledge of cell biology. It is possibly because of this that virology is rarely taught in schools.

The books described in this section represent a comprehensive but not exhaustive selection of the most commonly used books in the field published, with a few exceptions, between the years 1970 and 1981. For convenience they are divided into 15 broad classes although inevitably, in such an interdisciplinary field, there is much overlapping. The classes, and the number of books listed (in brackets), are given below. The full bibliographic details of these books are listed on pp. 178–188 and a selection of them are reviewed on the following pages.

GENERAL (10)
HISTORY (7)
BACTERIOPHAGES (10)
CHEMOTHERAPY (including interferon) (12)
IMMUNOLOGY (7)
INVERTEBRATE VIROLOGY (4)
MEDICAL VIROLOGY (11)
METHODOLOGY AND DIAGNOSIS (16)
MOLECULAR VIROLOGY (15)
PLANT VIROLOGY (including fungal viruses) (10)
SLOW VIRUSES AND VIROIDS (4)
SPECIFIC GROUPS (17)
STRUCTURE AND MORPHOLOGY (6)
TAXONOMY AND CLASSIFICATION (2, plus 1 report)
TUMOUR VIROLOGY (18)
VETERINARY VIROLOGY (including zoonoses) (8)

Under the heading of 'GENERAL' will be found books dealing with the broad subject area: animal (including man), plant and bacterial virology. Additionally, because plant virology is not so extensively covered by the literature, books covering only animal and bacterial viruses will also be found here. Texts covering the fundamental aspects of viruses such as biochemistry and genetics will be found under 'MOLECULAR VIROLOGY'. Finally, those dealing with just one virus or a group of viruses are included in the class 'SPECIFIC GROUPS'.

The relative representation of books within each subject shows which areas of virology have been of most concern to scientists this past decade. Probably the most striking thing in this regard is the great interest shown in TUMOUR VIROLOGY, which with 18 titles boasts the largest population of books. To some extent this reflects the massive funds that have been committed to the cause of establishing a link between viruses and cancer, particularly in the USA. One may expect to see a levelling off of interest in the near future. As our knowledge of individual viruses has increased greatly, review articles have been encapsulated in books to obtain wider circulation, and this is reflected in the high total of textbooks under the label 'SPECIFIC GROUPS'. The need for practical guides to the subject has also increased as the number of workers, particularly in hospitals, has increased.

A point made earlier about the widespread use of microbiology textbooks to teach virology at undergraduate level is illustrated by the relatively few books published under the headings of 'GENERAL', 'MEDICAL VIROLOGY' and 'VETERINARY VIROLOGY'.

Books devoted to MOLECULAR VIROLOGY are fairly numerous. It should be noted that many of the general texts and those devoted to individual or groups of viruses also provide extensive coverage.

It may appear that IMMUNOLOGY is poorly represented but this is in many ways due to the excellent coverage of virology in some of the pure immunology books. Finally, the emergent field of CHEMOTHERAPY boasts the not insubstantial total of 10 books, largely as a result of the considerable attention given to interferon as an antiviral agent. The recency of this interest is demonstrated by the fact that half of the books on chemotherapy have been published since 1980.

Individual Virology Books

General (bibliography entries 117–126)

It is not easy to cover the whole area of virology in just one volume (*see* the 18 volumes of *Comprehensive virology*, edited by Fraenkel-Conrat and Wagner (1974–80), entry 95). Books that attempt this feat usually do so by being highly selective or by emphasizing some aspects at the expense of others, such as plant virology.

General virology (3rd edn) by Luria *et al*. (1978) is notable for its treatment of virology as a whole with little regard for the divisions between bacterial, animal and plant viruses. Following the pattern of earlier editions, the first appearing in 1953, the introductory chapters deal with basic theory and practice of virus assay and properties of viruses, and this material then serves as the background to the central section of the book, dealing with virus growth cycles. The much revised third edition concentrates on biochemical and genetic aspects of virology, discussing the contribution of virus research to the understanding of macromolecular assembly and DNA restriction enzymes. The relationship of viruses to specific diseases and to cancer is not forgotten. This book continues to be one of the most popular set books for courses that contain a comprehensive virology component.

A competitor to *General virology* in the undergraduate sector of the book market is *The biology of animal viruses* (2nd edn) by Fenner *et al*. (1974). It covers the same ground as *General virology* and may be said to be more comprehensive and certainly larger. However it is becoming dated and does not compare so favourably price-wise. Just under half the book deals with the fundamental aspects of viruses, structure, function, multiplication, etc. The rest is concerned with viruses in relation to their animal hosts, immunity, interferon, pathogenesis, control and epidemiology. It is probably slanted more toward the medical and veterinary student.

More for the medical student but useful to all virology students is *Principles of animal virology* edited by Joklik (1980). It is divided into two sections: first the mandatory introductory chapters on basic virology, and secondly a section entitled 'Clinical virology', which discusses the major groups of viruses and includes a chapter on diagnostic virology. It has a pleasant, fresh format and plenty of references.

Books providing a somewhat more selective treatment of virology include two highly readable, up-to-date and fairly cheap volumes. *Introduction to modern virology* (2nd edn) by Primrose and Dimmock (1980) does not attempt to cover the entire field but emphasizes the biochemical and genetic aspects of virology. The second edition has been enlarged to take account of the interaction of animal viruses with their host at the level both of the cell and of the whole organism. *Introduction to virology* by Smith and Ritchie (1980) similarly concentrates on the biochemistry and genetics of viruses. In addition it has a chapter on the control of virus diseases which includes the use of viruses in the biological control of pests.

Viruses of vertebrates (4th edn) by Andrewes, Pereira and Wildy (1978) is unlike any of the books treated so far in that it deals systematically with the viruses themselves and their relationships with each other. The pathogenic effects are considered only so far as is necessary to identify a virus and to indicate its importance. The treatment of the virus families and their members is uniform: morphology, chemical composition, physicochemical characters, cultivation, habitats, etc. Major reviews are listed and an extensive bibliography follows each virus family. Whilst not cover-to-cover reading, it is an essential reference tool for workers and advanced students.

History of Virology (bibliography entries 127–133)

Most virology textbooks precede their accounts of the virus groups and their diseases with a short history of the disease. Another source of historical information, although more scattered, is the extensive literature on the history of medicine. The Wellcome Foundation is a major funder of this type of work.

Two important and rather similar books plotting the development of the subject of virology itself, have appeared recently: *The virus: a history of the concept* by Hughes (1977) and *An introduction to the history of virology* by Waterson and Wilkinson (1978). Both attempt to cover the evolution of the early concept of the virus. Hughes's account is the more academic, being originally part of a thesis on the history of medicine. In addition it follows the development of the germ theory of disease and contains biographies of Ivanovsky and Beijerinck and has an extensive bibliography. Waterson and Wilkinson's is a more popular account and essential reading for anyone remotely interested in virology. It contains a bibliography and biographical notes on over 100 individual virologists, both living and dead.

A different approach is followed in *Selected papers on virology* edited by Hahon (1964), which presents 40 outstanding journal articles by workers such as Jenner, Pasteur, Beijerinck, Delbrück and Maramorosch. The papers, prefaced and annotated, demonstrate the rationale and methodology of these pioneers. It could be subtitlted 'Virology's greatest hits'. In a similar vein, *Phage and the origins of molecular biology* edited by Cairns, Stent and Watson (1966) contains a collection of essays-cum-stories on the development of molecular biology and the part played by bacteriophage research during the 1940s and 1950s. Apart from the authors, contributors include Delbrück, Lwoff, Hershey and Dulbecco.

Finally, for those preferring a more light-hearted approach with an emphasis on human interest, there are *Virus hunters: the lives and triumphs of great modern medical pioneers* by Williams (1960) and *Fighting the unseen: the story of viruses* by Reidman (1967).

Bacteriophages (bibliography entries 134–143)

Phages are covered in many general texts, where they are extensively used to illustrate many of the genetic and biochemical characteristics of viruses.

The first book devoted solely to a single bacteriophage species was *The bacteriophage lambda* edited by Hershey (1971). Whilst a little old now, this book, in addition to providing a comprehensive description of the virological and molecular genetics aspects of lambda, still acts as a reference source for important areas of molecular biology.

Two classic texts that provide an account of the historical development of the subject, although somewhat out of date, are *Bacteriophages* by Adams (1959) and *Molecular biology of bacterial viruses* by Stent (1963).

More recently *RNA phages* edited by Zinder (1975) and *Single-stranded DNA phages* edited by Denhardt (1978), both published by the Cold Spring Harbor Laboratory, demonstrate how a concerted effort on these relatively simple viruses has resulted in insights into fundamental genetic processes. They are intended for specialists and non-specialists alike, but particularly for advanced students.

Unusually for this field, *Bacteriophages* by Douglas (1975) is a short and readable account of bacteriophage biology, aimed largely at the intelligent lay reader, older school-children and first-year undergraduates. It is good on basic techniques and it has a strong and lengthy chapter on phage genetics.

Chemotherapy (Including Interferon) (bibliography entries 144–155)

Control of viral infections is usually achieved by prevaccination of susceptible individuals, but lately interest has resurfaced in the chemotherapy of disease. A number of books on both areas have been published recently.

Looking at the broad area, *Developments in antiviral therapy* by Collier and Oxford (1980) reviews recent research in this field. A similar book is *Specific treatment of virus diseases* edited by Bauer (1977).

More specific attention to individual viral disease control is given in *Chemotherapy and control of influenza* edited by Oxford and Williams (1976) and *Chemotherapy of herpes simplex virus infection* by Oxford *et al.* (1971).

Ribavirin — a broad spectrum antiviral agent by Smith and Kirkpatrick (1980) deserves mention as it is one of the first books devoted to a single antiviral drug.

Set to become the standard reference text in the study of interferon is the three-volume work edited by Gresser (1979–82) and simply called *Interferon 1 1979*, *Interferon 1980* and *Interferon 1981*. Other books on this subject are *Effects of interferon on cells, viruses and the immune system* edited by Geraldes (1975) and *The biology of the interferon system* by de Maeyer (1981).

All these books are really intended for workers and practitioners in hospitals, although those on interferon may have a wider appeal.

Immunology (bibliography entries 156–162)

On the whole, books on viral immunology tend to be heavy going for anyone not actively involved in the field. For an introductory account it is better to tackle *Essential immunology* (4th edn) by Roitt (1980) or to read an account in a general virology textbook.

Viruses and immunity by Koprowski and Koprowski (1975) is a fairly compact book that provides an introduction, albeit a complex one, to the basic immunopathology of viral infections, the relationships of viruses to certain cancers and auto-immune diseases. It also includes a speculative chapter on future vaccines. In a similar mould is *Viral immunology and immunopathology* by Notkins (1975). It begins with the history of viral immunology and proceeds to cover the synthesis and properties of viral antigens, humoral and cellular immunity, virus neutralization, the role of inflammatory cells and effector molecules in combating viral infection, and the genetic control of resistance. Hepatitis, leukaemia and Epstein–Barr virus are singled out for special attention.

Aimed at workers in diagnostic laboratories of hospitals as well as research laboratories is *Viral immunodiagnosis* by Kurstak and Morisset (1974). It deals with a large range of viruses, mainly of medical importance, from the point of view of diagnosis and basic research. Comparative descriptions of various immunodiagnostic techniques for herpes, influenza, rabies, rubella, hepatitis, and cancer viruses are also covered.

A unique publication is *Vaccine preparation techniques* edited by Duffy (1980). It is data-based and contains vaccine preparation techniques, mainly viral, derived from US patents issued since 1976. It has detailed technical information on the preparation of human and animal antiviral vaccines including hepatitis A and B, influenza, varicella, cytomegalovirus, rabies, foot-and-mouth disease, etc. Short descriptions of a number of immuno-stimulants such as interferon are provided, and an index of diseases, commercial companies, inventors and US patent numbers completes the work.

Invertebrate Virology (bibliography entries 163–166)

Invertebrate virology has not received much attention from book publishers, although this may change with the use of these viruses in the biological control of insects, etc.

Virus–insect relationships by Smith (1976) is a comprehensive but concise account of insect viruses. The first half is taken up with a description of the different types of virus and the diseases they cause. The second half includes chapters on infection, replication, serology and the mass rearing of insects for the study of these viruses. It comprises an introduction to the subject and succeeds Smith's earlier work *Insect virology* (1967).

Two more weighty contributions are *Viruses and invertebrates* edited by Gibbs (1973) and the specialized *Aphids as virus vectors* edited by Harris and Maramorosch (1977).

Medical Virology (bibliography entries 167–177)

Quite a few books serve the general field of medical virology, including many of the texts covering the whole of microbiology such as *Microbiology* (3rd edn) by Davis (1978) and *Medical microbiology* (13th edn) by Duguid, Marmion and Swain

(1978). But of those covering virology alone, *Medical virology* (2nd edn) by Fenner and White (1976) is the most widely used by undergraduates. The first half of the book concentrates on the principles of animal virology, classification, cultivation, immunology, epidemiology and diagnosis, while the second concentrates on specific groups of viruses, including 'slow viruses'. The developments in molecular virology that took place between 1970 and 1975 are also covered, with particular emphasis on the tumour viruses.

A much briefer account of viral infections of man is contained in *Essentials of medical virology* by Pumper and Yamashiroya (1975). It is a compilation of material on basic and clinical aspects of the subject and contains short notes on all major human viruses in a stereotyped fashion: propagation, incubation period, pathogenesis, pathology, prognosis, diagnosis, etc. It is designed as a handbook for medical students and practitioners.

Not attempting to cover the entire field, *Recent advances in clinical virology* edited by Waterson (1977) contains 12 reviews on human viral diseases, half of them being concerned in some way with the diseases of the nervous system. Topics covered include herpes encephalitis, papovaviruses, Lassa fever, rubella and measles vaccines. Scrapie for some reason is also included, this being a virus of lesser animals.

Viral infections of humans: epidemiology and control edited by Evan (1976) is a collection of edited papers on human viral infections. The emphasis is on the concepts and methods of epidemiology and surveillance and also the major groups of viral pathogens and diseases such as cervical cancer, Burkitt's lymphomas and nasopharyngeal carcinoma.

Methodology and Diagnosis (bibliography entries 178–193)

Virology, like all the biological sciences, is a practical subject. Whether in a research laboratory or a diagnostic laboratory the need for descriptions of techniques is paramount. Most workers have a favourite methodology book and it is from these, not usually from research papers, that their knowledge of techniques is derived.

The most comprehensive treatment of methodology is undoubtedly the six volumes of *Methods in virology* edited by Maramorosch and Koprowski (1967–). For details of this *see* p. 91.

Comparative diagnosis of viral diseases edited by Kurstak and Kurstak (1977–81) in four volumes, is set to become the standard work in the area of viral diagnosis. Volumes 1 and 2, jointly subtitled *Human and related viruses*, deal with DNA and RNA viruses respectively. Similarly volumes 3 and 4, subtitled *Vertebrate animal and related viruses*, cover DNA and RNA viruses of animals.

One of the most widely used laboratory manuals is *Diagnostic procedures for viral, rickettsial and chlamydial infections* edited by Lennette and Schmidt (1979), now in its 5th edition and containing over 1100 pages. The book is divided into two parts. The first concerns the general principles underlying the laboratory diagnosis of viral infection, prevention of laboratory infection and details of techniques such as cell culture, fluorescent antibody, immunoenzyme and other immunological methods. The second part covers 24 viruses or virus groups mainly from a disease point of view. A more concise coverage of the same material is given by *Diagnostic methods in clinical virology* (3rd edn) by Grist *et al.* (1979).

Cell culture has become an essential part of virus studies and vaccine production. *Cell and tissue culture* (5th edn) by Paul (1975) is a vital reference tool for this work. As well as detailed instructions as to how to prepare the cultures, there are some very useful sections entitled 'Tissue culture in biomedical research' and 'Virology and host–parasite relationships and cancer'. It is easily comprehended by students and researchers alike. In a similar area, *Biochemical methods in cell culture and virology* by Kuchler (1977) summarizes methods of growing, handling and studying animal cells and viruses.

Molecular Virology (bibliography entries 194–208)

A great deal of research has been done in virology in recent years using the methods of molecular biology and this is reflected in the literature. In fact it is probably true to say that most books published in virology since the mid-1970s have emphasized this element. Books dealt with here are those that provide an overall picture of the nature of viruses and their interaction with the host at the molecular level.

A good introduction to the subject is given in *The biochemistry of viruses* by Martin (1978). It is a concise book, omitting epidemiological and pathological problems, that compares the variety of structures found in viruses and the different strategies that viruses adopt when they infect cells. Chapters on quantitative methods, architecture of viruses and an interesting one entitled 'Viruses and the biosphere' make up a book that is useful to undergraduates of all biological persuasions.

Although of the same name, *The biochemistry of viruses* edited by Levy (1969) is older, more comprehensive, more advanced and thus aimed at a different market: the research scientist or student. The latter should find the two-volume *The molecular biology of animal viruses* edited by Prosad-Nayak (1977) a more up-to-date reference source. The first two chapters are on 'Symmetry in virus architecture' and 'Interferon' respectively; the remainder deal systematically with the molecular biology of the major groups of animal viruses and contain many references.

Dealing with molecular biology of viruses more succinctly are the 64 pages of *Molecular virology* by Pennington and Ritchie (1975) and the slightly longer *Molecular virology* by Knight (1974).

The genetics of bacteria and their viruses by Hayes (1976) is a popular set book for many microbiology courses. It is fairly advanced and covers the relatively recent developments in genetics and molecular biology. It contains over 80 pages of references.

Host–virus interactions are the subject of *Virus receptors; Part I, Bacterial viruses* edited by Randall and Philipson (1980) and *Virus receptors; Part II, Animal viruses* edited by Lonberg-Holm (1980). These books are obviously aimed at the specialist, who will appreciate the up-to-date subject matter. On the same theme, *Cell membranes and viral envelopes* edited by Blough and Tiffany (1980) presents the state of the art in two very expensive volumes.

Plant Virology (Including Fungal Viruses) (bibliography entries 209–218)

Knowledge of plant viruses is limited and scattered, as far as the published works are concerned. Information on plant viruses appears as plant pathology, phytopathology and plant diseases as well as invertebrate virology; insects and others are often very important vectors of plant disease.

Recently published and set to become the definitive work in this area is *Handbook of plant virus infections: comparative diagnosis* edited by Kurstak (1981). It provides, in 27 chapters, information on unifying concepts of comparative virology; a detailed and comprehensive treatise of infections; and new information and a look at the direction of the latest research. It is intended for both the field worker and laboratory scientist.

Plant virology: the principles by Gibbs and Harrison (1979) is intended as an introduction to the subject. It is essentially divided into four parts. The first deals with the main groups of viruses with thumbnail sketches of the most important. Classification and characterization are covered next. The third part describes the viruses' transmission and ecology whilst the last part is devoted to viruses of organisms other than higher plants. The book contains over 1000 references.

Covering similar ground is *Plant viruses* by Smith (1977), its popularity demonstrated by the fact that it is now in its 6th edition. It also covers viruses of algae, fungi and mycoplasmas. *Plant virology* by Matthews (1976) is a reference tool, containing some 800 pages, that is primarily aimed at the active research worker.

Symposium on fungal viruses edited by Molitoris *et al.* (1979) presents the proceedings of a symposium on viruses of fungi or mycoviruses. It includes abstracts of papers on extrachromosomal vectors in fungi and abstracts of posters on the fungal viruses, with chapters on evolution, characterization, methods and taxonomy.

Slow Viruses and Viroids (bibliography entries 219–222)

Interest has grown in the diseases caused by these viruses or virus-like organisms.

Viroids and viroid diseases by Diener (1979) begins with a historical account of the discovery of viroids, then deals with their natural occurrence and role as plant pathogens. Other sections consider biological tests and in-vitro procedures, their mode of replication and specific examples such as kuru and scrapie.

Based on an international symposium held in the USA in 1978, *Slow transmissible diseases of the nervous system* edited by Prusiner and Hadlow (1979) consists of two fairly lengthy and complex volumes. The first covers the clinical, epidemiological, genetic and pathological effects of the spongiform encephalopathies, whilst the second is devoted to pathogenesis, immunology, virology and molecular biology of the same.

For those requiring a gentle lead-in to the last book, *Slow viruses* by Adams and Bell (1976) provides a suitable introduction, as does *Slow virus diseases of animals and man* by Kimberlin (1976).

Specific Groups (bibliography entries 223–239)

It would have seemed inconceivable forty, or even thirty years ago that enough would be known about one virus or even a family of viruses to fill one complete book. However, today there are many books — usually compiled by a host of specialists — published on individual viruses.

The herpesviruses edited by Kaplan (1974) contains chapters by 28 internationally known virologists. Topics include: history and classification, fundamental properties, immunological relationships and the major members of this important group such as Epstein–Barr virus, cytomegalovirus and herpes simplex virus. Two chapters on chemotherapy complete this standard reference text. Because of recent developments a new edition is probably called for.

Spanning three volumes, *Rhabdoviruses* edited by Bishop (1981) is another multi-author compilation. Volume 1 deals with the natural history, methodology, biochemical composition and structure of rhabdoviruses. The second volume concerns itself more specifically with the replication and genetic aspects whilst the third volume covers interferon, vaccination and chemotherapy and specific rhabdoviruses of reptiles, fish and plants. The two most important members, vesicular stomatitis virus and rabies, are cited throughout this mini-series. At just over £90 for the set, libraries will probably be the major buyers.

The togaviruses: biology, structure and replication edited by Schlesinger (1980) comprises 21 chapters that examine the biology, structural and biochemical characteristics of all viruses assigned to the family Togaviridae, which include the arthropod-borne alpha and flaviviruses and the 'non-arbo' togaviruses of the genera *Rubivirus, Pestivirus* plus unclassified members. Entomologists may find this volume useful, as will the togavirus specialist.

Not as comprehensive as the last title, *Non-arthropod-borne togaviruses* by Horzinek (1981) concentrates on those viruses not transmitted by insects and the like.

The influenza viruses and influenza edited by Kilbourne (1975) covers the basic virology of these viruses with chapters on virus isolation, propagation, animal host range, antigenic variation, immunology, influenza in man and animals and epidemiology. An update on developments in this field may be obtained from *Influenza: a Royal Society discussion*, organized and edited by Tyrell and Pereira (1980).

Replication of mammalian parvoviruses edited by Ward and Tattersal (1978) is not as narrow in subject material as the title suggests, being a comprehensive collection of reviews and original research articles covering mammalian parvoviruses including canine parvovirus and feline panleukopaenia virus.

The published proceedings of a summer school entitled *The molecular biology of picornaviruses* edited by Perez-Bercoff (1979) examines the properties and behaviour of this very important group, having members such as poliomyelitis and foot-and-mouth disease virus. Capsid structure and assembly, viral protein synthesis and interferon are some of the topics discussed.

Rabies has attracted much attention in the form of books, some simple and others complex. *Rabies: the facts* by Kaplan and others (1977) is a fairly non-technical account of the biology and medical details of rabies and can be read by virologists and non-virologists alike. A more detailed and comprehensive reference book is *The natural history of rabies* edited by Baer (1975). This two-volume treatise covers the history of the disease, detailed description of the virus, pathology, diagnostic techniques, epidemiology and control. Students and practitioners of veterinary and human medicine will most benefit from this work.

Structure and Morphology (bibliography entries 240–245)

On the whole, books describing structure and morphology of viruses with their many large electron micrographs resemble atlases. One such book is *Virus structure* by Horne (1974), measuring 20 × 25 cm. It provides an introduction to the study of viruses with particular reference to structure and morphology as determined by electron microscopy. After a general chapter on symmetry, small and double-stranded DNA icosahedral viruses, helical viruses and those with a more complex symmetry are discussed in more detail. *An electron microscopic atlas of viruses* by Williams and Fisher (1974) has less information but makes up for this in photographs, having some 31 plates showing a variety of viruses with short descriptions and an indication of (not so) recent work on its structure. A similar text is *An atlas of insect and plant viruses* edited by Maramorosch (1977), which is a large and expensive book, also covering mycoplasma viruses and viroids. The full-size electron micrographs are accompanied by brief descriptions of the viruses.

Containing more detailed information is *Ultrastructure of animal viruses and bacteriophages—an atlas* edited by Dalton and Haguenau (1973). Each chapter deals with a specific group of viruses and details the ultrastructural aspects of viral replication and the internal and external organization of virions as well as including a brief summary of biochemical and immunological information on the groups.

A short, concise and inexpensive account of virus structure is given in *Structure and function of viruses* by Horne (1978). This book is meant as an introduction to virology and briefly covers growth cycles of viruses as well as structure: it can be read by older school-children or undergraduates. In a similar vein but more extensive is *Virus morphology* by Madely (1972).

Taxonomy and Classification (bibliography entries 246–248)

Most general textbooks have a chapter on classification; the more recent the book the better the chapter, is a general rule.

The best presentation, however, comes from the horse's mouth as it were. *Classification and nomenclature of viruses* edited by Matthews (1979) is the third report of the International Committee on Taxonomy of Viruses (ICTV). The main section summarizes the state of approved virus nomenclature and taxonomy as it was following the plenary meeting of the ICTV at The Hague in 1978. It is divided into four parts: 'The viruses', 'Presentation', 'Virus diagrams', and 'The families and groups'. The report also contains a list of officers and members of the ICTV. The first report incidentally is of interest because it charts the history of classification in virology.

One of the major problems in viral classification has been the unification of the interests of the plant virologists with those of animal virologists and the others. For this reason plant virologists may prefer *Contributions to the systematic plant virology* edited by Hanson in two volumes. Volume 2 has its own title: *Codes, data and taxonomy*.

Tumour Virology (bibliography entries 249–267)

Tumour virology has witnessed a great growth of interest over the past decade with the net result that there is a wide choice of books, many with confusingly similar titles, to choose from. In general these books are addressed at the specialist.

For many years *Oncogenic viruses* (2nd edn) by Gross (1970) was the definitive authority. Although still a classic, the rapid developments that followed have made it out of date.

Two mammoth Cold Spring Harbor Publications, *DNA tumor viruses* (2nd edn) edited by Tooze (1981) and *RNA tumor viruses* edited by Weiss *et al.* (1981), have tended to make many similar, but older, texts redundant. The first describes the cellular microbiology of SV_{40}, polyomavirus, adenoviruses, papilloma viruses and the herpesviruses with particular emphasis on their structure and composition, replication and their role in transformation. The second presents the concepts of retrovirology, including an extensive catalogue of virus strain mutants, endogenous proviruses, restriction maps and rules for nomenclature. The many research workers in cancer institutes and molecular biology laboratories will benefit most from these two books.

In very similar territory, *Molecular biology of RNA tumor viruses* edited by Stephenson (1980) is also aimed at research workers. After a chapter on the history of tumour viruses in general, it reviews the endogenous retroviruses, their transmission, nature and origin of the transforming RNA viruses and their translational products such as reverse transcriptase. The type B and D retroviruses are also covered. Copious references are provided at the end of each chapter.

Selected papers in tumor virology edited by Tooze (1974) is a collection of articles important to the field, and includes landmark papers describing the origin of tumour virology, structure, genetics and replication of polyoma SV_{40} and other implicated tumour viruses.

Veterinary Virology (Including Zoonoses) (bibliography entries 268–276)

There are many books on animal virology, although only a small number deal specifically with viruses of domestic and farm animals. Normally such viruses and their diseases will be found scattered around in veterinary textbooks or in books dealing with the diseases of one particular animal. Two examples of this are *Hagan and Bruner's Infectious diseases of domestic animals* (7th edn) edited by Gillespie and Timoney (1981) and *Diseases of poultry* (7th edn) edited by Hofstad *et al.* (1978), which contain excellent sections on viruses of animals and poultry respectively.

The most comprehensive and comparative account of veterinary virology can be found in *Animal microbiology*; vol. 2, *Rickettsias and viruses* by Buxton and Fraser (1977). It covers the basic properties of animal viruses, their classification, the disease, diagnosis and prophylaxis. The emphasis is of course on viruses of farm and domestic animals but human and fish viruses are also dealt with. The major groups of RNA and DNA viruses are treated systematically: history and distribution, morphology, physicochemical properties, antigenic properties, cultivation, pathogenicity, diagnosis, control, etc. The book is of use to students and vets.

Veterinary virology by Mohanty and Dutta (1981) is a more concise and very up-to-date account of the viruses of domestic, farm and wild animals. It is divided into two parts: 'General animal virology' and 'Viruses of animals'. It is the second part that will be of most interest to veterinary undergraduates and the veterinary profession. It groups the viruses under the animal species, such as bovine, equine, ovine, avian, etc. Chapters on viruses of laboratory, wild and zoo animals and viral zoonoses are included.

4.4 CONFERENCE PROCEEDINGS

We have seen in Chapter 3 that the published proceedings can never fully represent all that took place at a conference. Nevertheless, they must be considered, since they form part of virology's literature (*see* bibliography entries 19–31). The published proceedings, whether in book, journal or report form, may consist of all or just a selection of the papers presented. Sometimes these are reproduced in full, but they may be edited or reduced to abstract form. The discussions and resolutions that accompany each paper may also be reported, either verbatim or, more normally, in a condensed form.

The real value of conference proceedings is in some doubt, many seeing them as rather half-way-house communications with little permanent or archival value. Certainly the quality of the papers — rarely subjected to refereeing — may leave much to be desired. Such doubts may well explain why, despite the high value placed on conferences by scientists, the published proceedings are relatively little used. However, more likely the explanation lies with the late appearance and general inaccessibility of the proceedings. Commonly proceedings appear as much as a year or two after the conference was held; indeed, these papers are the lucky ones for it has been estimated that as many as half the papers given are never published at all. Thus to the best of our knowledge no proceedings have ever emerged from the four bi-annual Rhabdovirus Symposia, which began in 1973.

Difficulties do not end as a result of the publication of a proceedings for there is still what seems to be the almost insuperable problem of tracing it and locating it, a question considered more fully in section 5.6. For a start, few libraries maintain comprehensive collections (though borrowing through the British Library Lending Division, who maintain the most complete collection of proceedings in the world, might help here) and those that do, often, confusingly, store them separately from other documents on the same topic in some remote or inaccessible store. Allied to this is the fact that the proceedings may be published as part of a journal, in which case it would be somewhere else again in the library.

Undoubtedly though, the greatest difficulties are created by the variable way in which conferences are cited — no one seems to know or agree how best to describe them. Thus they may variously be entered under: the official name of the meeting, the sponsoring body (frequently bodies), editor, place held, title (and because of the complex nature of the title there may be differing ways of specifying this), keyword, and authors of individual papers (though this form of analytical entry is regrettably becoming rare). As an example take the following conference: the 10th Federation of European Biochemical Societies Meeting, Paris, 1976. The title was *Organization and expression of the viral genome*, and the editors were F. Chapeuille and G. Grunberg. The proceedings were published in 1977 by North-Holland (Oxford) as volume 39 of *FEBS Proceedings of Meetings*. This may be cited variously under the name of the Federation, the editors, the series title (*FEBS Proceedings of Meetings*) and various keywords. Few reference tools cater for all approaches, the *British National Bibliography* being perhaps a notable exception.

Thus unless one has a sufficiently detailed knowledge of the conference sought — and this must be the exception rather than the rule — to check out all the possible entry permutations (plus time and patience of course), a truly effective search for a proceedings cannot be mounted.

A number of journals take it upon themselves to publish the abstracts and sometimes the full texts of papers given at conferences, usually conferences held by the journal's sponsor/publisher (normally a society). The *Society for General Microbiology Quarterly* and the American Society of Microbiology's *Microbiology* are perhaps the best examples in virology. The *Quarterly* contains abstracts of papers read at ordinary meetings and at many of the Society's specialist group meetings, including the Virus Group. *Microbiology* is an annual publication containing the proceedings of conferences organized by the ASM the previous year. The most recent edition (1981), for instance, contains abstracts of papers on the following topics of interest for virologists: gene movements and laboratory evolution, including phage and viral specific recombination; transposons as tools in microbial genetics and molecular evolution; infectious diseases — describing recently recognized infectious agents; regulatory mechanisms; and finally diagnostic immunology, genetics and molecular biology of industrial micro-organisms.

A journal that, whilst not as consistently relevant as the two above, does deserve mention is *Federation Proceedings*, which contains articles representing papers given at symposia organized by the Federation of American Societies for Experimental Biology or one of its member societies; approximately 10% of the journal's contents are concerned with virology, usually discussed under the heading of 'Immunological mechanisms of virus disorders' or 'Tumour antigens'.

Finally three other journals are occasional but important carriers of conference proceedings: *Proceedings of the Society for Experimental Biology and Medicine*; *Proceedings of the Royal Society of London; Series B, Biological Sciences*; and *Proceedings of the National Academy of Sciences of the USA: Biological Sciences*.

4.5 DISSERTATIONS

Dissertations, theses — call them what you will — do play an important, if specialist, role in the virology information system. Their value lies mainly in the level of detail provided, particularly in regard to the methods adopted. Whereas the practical problems and real 'nitty gritty' associated with the conduct of the experiment (at its base level, the washing of laboratory instruments) are generally skated over in research papers they are given full airing in the dissertation. Thus in many ways dissertations can be used as a manual of practice. They are valued too for the supposed originality of the material, but this is a much more contentious issue. They are also likely to have the edge on journals when it comes to currency of reporting.

Their detractors would argue however that their sheer bulk, variability in quality and ponderous style make them an information source of last resort. Certainly the fact that few have indexes makes it difficult to retrieve information from them quickly. Nonetheless, the great advances that have been made in making dissertations more accessible will inevitably mean that they will become an increasingly important information source for virologists.

Section 5.7 deals with ways of tracing and locating dissertations.

4.6 REFERENCE WORKS

Encyclopedias and Handbooks (*see* bibliography entries 277–282)

Encyclopedias and handbooks are called quick reference tools because they provide a 'facts at your fingertips' approach to a vast store of data. This store represents in essence the basic wisdom of a subject (as in the case of a 'Handbook of microbiology') or the universe (as in the case of *Encyclopaedia Britannica*). These tools are designed in such a way as to enable the quick retrieval of concise capsules of information on a multitude of topics by a wide variety of users. Combining the functions of dictionary, book, biography, review and yearbook as they do, they all hope to attract readers because of the appeal of one-place reference (i.e. the reader goes to one source for a wide variety of information). The greatest difficulty encountered by quick reference tools is in keeping abreast of the rapid advance of knowledge; revision tends to be irregular and too infrequent, with most running 7–10 years in arrears.

There can be few people indeed who have not turned to the *Encyclopaedia Britannica* to help write a last-minute essay, obtain a familiarity with an unfamiliar field of knowledge or to check the date of an important event. This 30-volume set (in the 15th edn) consists of the *Propaedia* (an outline of knowledge), the *Macropaedia* (the substantial account of subjects, events, people) and the *Micropaedia* (an index to the main articles, which also supplies packets of information with each index entry to satisfy the quick reference need).

The virology entry, written as it is by W. C. Summers of Yale University, is undoubtedly authoritative. Whilst being concise and informative it is also, at 14 000 words in length, far from being insubstantial. The article is very much addressed to the biology undergraduate. A comprehensive account of virology is given, with the following topics being touched upon: history of viral diseases, morphology and chemistry, classification, viral growth and development, types of infectious process (lysogeny, transduction, etc.), viral genetics, cultivation, pathogenesis (tumour virology, etc.), and origin and evolution. A bibliography of 10 or so virology books (all unfortunately pre-1973) provides for the user requiring further information. Surprisingly, for a work that attempts alphabetically to compartmentalize information in over 30 volumes, there is little scatter of information in the case of virology. The only other entry of any note is 'Infectious diseases', written by A. B. Christie of Liverpool University, where general information on diseases such as routes and modes of infection, immunology, prevention, etc. is covered. Specific diseases such as smallpox, polio, rabies, hepatitis, influenza and rhinovirus infections are described.

Other encyclopedias, even such scientific ones as *Van Nostrand's scientific encyclopedia* (6th edn, 1983), and the *McGraw-Hill encyclopedia* (5th edn, 1982) and *Veterinary encyclopaedia* (1968) are of limited use.

Handbook of microbiology, edited by Laskin and Lechevalier (1973–4), is becoming a little outdated, although apparently a second edition is planned. According to the editors, this four-volume set, running to over 3500 pages in total, 'functions not as a treatise of systematic microbiology but rather a handbook where people will be able to look up information about micro-organisms'. Volume 1 is of the most interest to virology. Entitled *Organismic microbiology*, it covers the major groups of microbes including viruses, methodology and general reference data. The data include safety regulations, member countries of IAMS, rules of nomenclature, a literature guide to microbiology, and colleges and universities (in the USA) offering degree courses in microbiology.

After a brief introduction the division of virology is based on host affinity and disease:

Introduction to the systematics of viruses
Viruses of plants
Viruses of invertebrates
Viruses of vertebrates
Phage typing

Volume 2, *Microbial composition*, is mainly in tabular form. It covers base composition and buoyant densities of nucleic acids of phages. Volume 3, *Microbial products*, has a section on compounds that inhibit virus multiplication, with their structure, source, type of activities, etc. Volume 4, *Microbial metabolism, genetics and immunology*, under the heading of 'Genetics', has linkage maps of bacteriophages T_4 and $\phi X174$ and a genetic and molecular map of *E. coli* phage lambda. Processes such as transduction are also covered here. Under 'Immunology' the antigenic relationships of some animal viruses are included.

Biology data handbook (2nd edn), compiled and edited by Altman and Dittmer (1972) and published by the Federation of American Societies for Experimental Biology, is a three-volume treatise with only the first volume pertaining to virology. It covers genetics, cytology, reproduction, development and growth and includes tables giving the properties of biological substances and information about some of the many widely used materials and methods in biology. The generation times of a few viruses are given under the subheading 'Cell division frequency'. *Cell biology* is another Federation publication. It is edited and compiled by Altman and Katz (1976) and will be helpful to virologists in relation to cell culture and animal cell lines.

No handbooks or encyclopedias exist in name in virology although a number may be said to perform a similar function: *Viruses of vertebrates* (4th edn) by Andrewes, Pereira and Wildy (1978) provides a concise and easy reference source for information on animal viruses (*see* 'Books', section 3.3); *Essentials of medical virology* by Pumper and Yamashiroya (1975) acts as a handbook for medical students and practitioners and contains short notes on all the major human viruses.

Subject Dictionaries (*see* bibliography entries 283–290)

The function of a dictionary, whatever the subject field, is to define and standardize the working vocabulary of that subject. Whilst most general dictionaries are used primarily to check the spelling of words (and indeed this is one use of subject dictionaries also), most subject dictionaries are used to obtain quick understanding of subjects' concepts, processes, techniques and sometimes organizations and individuals. In some respects then, subject dictionaries resemble encyclopedias and may be regarded as mini-encyclopedias.

Commenting on a recently published microbiological dictionary, one reviewer in the *SGM Quarterly* (Nov. 1980) wrote rather harshly, 'In conclusion I cannot identify any market for this dictionary.' This nicely encapsulates the view that in such a rapidly developing area as virology, dictionaries can never be a very potent force. The argument continues that most of the relevant information is contained in relatively few advanced textbooks which most people interested in the subject possess already and that the items missing from these textbooks are most likely the more recent terms to appear in the literature — it is these very terms that are also likely to be missing from the dictionaries. This is undoubtedly true; dictionaries by their very nature cannot do justice to a complex and growing field unless continually updated; and most are lucky to be updated every five years. The lot of a dictionary compiler is not a happy one in a field such as virology that is plainly lacking in consensus, for even a good dictionary can be an easy prey for reviewers. They can criticize at length omissions or the balance of the contents, such as 'If DHSS merits an entry why not MAFF' or 'three and a half pages are devoted to microscopy but only a third of a page to continuous culture'.

Despite these criticisms however, the abundance of subject dictionaries in circulation provides ample evidence that they do meet a need. Perhaps the most pressing need stems from people from other fields who are attempting to come to grips with virology. Probably too their conciseness and 'facts at your fingertips' approach have a great deal to do with their appeal. And in such a multidisciplinary field as virology, dictionaries have the added attraction of bringing together in one place terms and topics that are normally scattered.

In the main the dictionaries that cover virology are aimed at the layman, science undergraduates, knowledgeable workers in related scientific fields and information specialists who want to break through the language barrier that separates them from the people they serve.

Until 1981 virology was served only by the dictionaries of the multifarious disciplines that either contain it, have an interest in it, or are related to it (i.e. biochemistry, genetics, microbiology, immunology, medicine and veterinary science — the emphasis in the latter two being firmly on the clinical signs, diagnosis and treatment of disease). The year 1981, however, saw the publication of a dictionary entirely devoted to virology, further establishing virology as a subject in its own right. Approximately 80% of the 2000 or so entries in *A dictionary of virology* by Rowson *et al*. (1981) describe individual or groups of viruses using names approved by or conforming to rules of the International Committee on Taxonomy of Viruses. Synonyms of the viruses are included in each main entry, under which their characteristics and taxonomic status (family, genus or species) are briefly set out. In most cases a leading recent reference to the literature is given to enable more detailed investigation. The remainder of the entries concern techniques, antiviral drugs and genetic, immunological and biochemical terms relevant to virology. A useful table on the taxonomy of vertebrate viruses is also contained within the work.

A dictionary of microbiology by Singleton and Sainsbury (1981) is an unrevised paperback version of the 1978 edition and covers terms, concepts, techniques, tests and other topics in bacteriology, mycology, virology and microbiological aspects of medicine, veterinary science and plant pathology. Major groups of viruses and some individual viruses get quite lengthy treatment. The nomenclature however is a little outdated, an example being 'leukoviruses'. The dictionary also includes a large number of cross-references and an appendix of microbial metabolic pathways.

A dictionary of microbial taxonomy by Cowan and Hill (1978) seeks to explain the concepts and procedures of modern microbial taxonomy. However it is mainly concerned with the classification of bacteria. The recent developments in viral taxonomy mean that this book is of limited value to virologists, although from a historical point of view brief biographies of leading microbial taxonomists may be of interest, as will chapters dealing with the history of the codes of nomenclature, the philosophy of taxonomy and source material for taxonomy.

Biochemistry pervades nearly all areas of virology and it is by far the least understood of the virological tools. *A dictionary of biochemistry* by Steneish (1975) contains over 12 000 entries taken from 200 reference books and research literature. Whilst perhaps virology is better served elsewhere, the concepts and techniques of biochemistry contained will be of great use as a source of reference to virologists.

A dictionary of biomedical acronyms by Dupayrat (1978) is rather a luxury but contains a number of useful acronyms found in the literature of virology such as c.p.e., CELO and RSV.

Because virology and molecular biology are almost indistinguishable at times, the *Glossary of molecular biology* by Evans (1974) is of relevance to virologists. The 400 or so entries include many terms used in virology. It is however badly in need of updating and enlarging. There are 139 references at the end of the book although none later than 1973.

Of the 6000 entries in *Glossary of genetics and cytogenetics* by Rieger et al. (1977), many deal with terms used in phage and animal virus genetics such as lysogeny, interferon, reverse transcriptase and virus maturation. There are also over 1000 references. It provides a comprehensive reference source to students of microbiology, to whom genetics is such an integral component of the course.

Like biochemistry and genetics, immunology is a vital ingredient of virology and therefore a knowledge of its language is essential. Amongst the 1300 or so entries in *A dictionary of immunology* by Herbot and Wilkinson (1977) is included a range of terms — directly and indirectly relevant to virology — concerning vaccination and serological tests such as the fluorescent antibody and immunodiffusion tests.

Biographical Dictionaries (*see* bibliography entries 291–297)

Personalities would not seem to merit the same attention in science as they do in the social sciences and humanities, where people are as much the object of study as their work (in fact the two are held to be inextricably entwined). Nevertheless there is a need for biographical sources in the sciences, especially for those that provide a subject approach and thus enable the identification of subject specialists, possibly the most important information sources of all. Other more mundane uses include: checking out an author's mailing address; establishing what are the achievements, educational and professional, of particular scientists; and, in a more bibliographic vein, finding out what various individuals have published.

To aid us in our criticisms of biographical works we assembled a list of virologists, some important and some rather less important, against which to test their coverage (*Table 4.5* provides details).

Found surprisingly in a monograph called *An introduction to the history of virology* by Waterson and Wilkinson (1978) is probably the most comprehensive listing of past virologists. Notes referring to over a hundred virologists of all nationalities, including a few living, may be found in the appendix to this work. The selection is rather eccentric at times, including for instance an Italian poet! Of our test sample of 29, 2 were mentioned.

Table 4.5 Coverage of virologists by biographical dictionaries

	American men and women of science	Who's who in British scientists	Introduction to the history of virology	Who's who
Almeida, J.				
Andrewes, C. H.			✓	✓
Baltimore, D.	✓			✓
Bishop, D. H.	✓			
Brown, F.				
Burke, D. C.		✓		
Deiner, T. O.				
Epstein, M. A.		✓		✓
Fenner, F.				✓
Fraenkel-Conrat, H.	✓		✓	
Horne, R. W.				
Kaplan, A. S.		✓		
Kaplan, C.	✓			
Koprowski, H.				
Kurstak, E.				
Laver, W. G.				
Lennette, E. H.				
Maramorosch, K.	✓			
Melnick, J. C.	✓			
Pereira, H. C.				✓
Pereira, M. S.				✓

(Table 4.5 continued)

	American men and women of science	Who's who in British scientists	Introduction to the history of virology	Who's who
Rhodes, A. J.				
Roizmann, B.	✓			
Schlesinger, R. W.	✓			
Smith, K. O.	✓			
Tinsley, T. W.				
Wildy, P.		✓		✓
Zhdanov, V. M.				
Zuckerman, A. J.		✓		

American men and women of science: physical and biological sciences (1979) is an authoritative and impressive source not just for American virologists but also virologists of other nationalities who have spent some time in the US either studying, researching or working. This qualification admits many foreign virologists. Although 130 500 scientists of some standing in the scientific community are listed alphabetically by name in this eight-volume work, from the subject index it is possible to identify about 800 virologists, including 9 of our sample. The entries are quite detailed, furnishing information about an individual's education, professional experience, research, field of interest and membership of societies. In addition the normal personal information is also supplied: age, birth, nationality and mailing address. (*American men and women of science* is also available on-line via DIALOG.)

Who's who in British scientists 1980/81 (1980) is the British counterpart, although nowhere near as comprehensive. Much the same information is presented as in the former publication but the entry is not as detailed. An essential difference is that the British work attempts to list a selection of the biographee's publications. A useful appendix lists the names and addresses of research establishments and scientific societies. No subject index is provided so it is not possible to identify virologists unless their names are already known. However a subject index is planned for the next edition of the work. Just 5 of our sample of virologists were traced in this work.

Probably the most widely available biography is the general *Who's who*, so it is interesting to see whether virologists are regarded highly enough to be considered. Not surprisingly, few are included; in fact just 2 of our test sample.

5 Searching the literature

5.1	**Guides to the literature**	115
5.2	**Tracing and locating journals**	117
5.3	**Tracing and locating periodical articles**	120
	Abstracting and indexing services	120
	Individual abstracting services	121
	Comparative assessment of the abstracting and indexing services most commonly used by virologists	133
	Current-awareness publications	134
5.4	**Subject bibliographies**	135
	Animal viruses	138
	Bacteriophages	138
	Human viruses	138
	Methodology	139
	Plant viruses	139
5.5	**Tracing and locating virology books**	139
5.6	**Tracing and locating conference proceedings**	143
5.7	**Tracing and locating dissertations**	145
5.8	**Classification schemes: finding virology documents in libraries and bibliographies**	146
	Dewey Decimal Classification	149
	Universal Decimal Classification	152
	Library of Congress scheme	154
	Barnard Classification	156
	National Library of Medicine scheme	157
5.9	**Epilogue**	157

5.1 GUIDES TO THE LITERATURE (*see* bibliography entries 298–304)

Today, because of the sheer amount of information there is about — and it is increasing by leaps and bounds — guides to the literature are virtually an indispensable tool for anyone whose job, research or study requires him or her to seek out information. In virology this must include everyone. The larger or more complex the field (virology is a member of the latter category) or the newer one is to a field (i.e. students), the greater the need to be guided through the information jungle. Indeed, it is a very rare person who can boldly assert that he is aware of all the information sources that might be of relevance to him. Our user survey indicated that it was only the very active researchers who felt at all secure in their knowledge, and only then within their narrow specialism and with the added qualification that this was true only 'as far as I know'.

If one sees the guides' function as simply helping people through the information labryinth, then why are they not more popular or widely available? Part of the explanation lies in the fact that many are poorly produced, showing no understanding of user needs; part lies with their limited horizons — either they are produced by librarians for librarians or, at the very least, for the tiny body of people who are positively enthusiastic towards information; and part rests with their failure to address themselves to the fields in which people actually work (i.e. one cannot expect a virologist to identify with, or find much of interest in, a guide covering the whole of the life sciences).

There is, however, possibly one further, more serious reason: the value of being well informed is not generally recognized. Many researchers for instance still pay allegiance to the practice of serendipity (the happy discovery of things by accident) whilst others allege that they have managed quite nicely to date, so why embroil themselves now in an activity they care little for anyway. The flaw in these arguments is of course that today, when competition is fierce, technological change rapid and government intervention in science and industry increasing (all factors that call for greater intelligence), we can no longer solely rely on the fruits of accidental discovery.

There are no guides specifically covering the virology literature. The closest we come is to guides covering the fields of medicine, veterinary science and biology — plainly some way from virology. *Comparative and veterinary medicine: a guide to the resource literature* (Kerker and Murphey) is probably the most relevant of these. The book is divided into four parts, the first covering indexing and abstracting journals, bibliographies, serials, reference works, handbooks and manuals; the second specific disciplines such as microbiology and genetics; the third veterinary medicine; and the fourth laboratory animals. Under the chapter on specific disciplines can be found an annotated list of 63 virology books, serials, and proceedings. Additionally, some virology journals, monographs and review serials can be located at a listing of periodicals of veterinary interest. The age of the book is a severe handicap to its use for much has happened in virology in the ten years since it last appeared (1973). The same can be said of another guide, *Biological and biomedical resource literature* (Kerker and Murphey), which suffers even more from having been published as long ago as 1968. The chapter on 'Materials of specialized interest' is where virology is dealt with, albeit rather summarily, with just 29 virology books being listed.

One of the more up-to-date accounts of literature is provided by *The use of medical literature* (Morton), published last in 1977 (2nd edn). Inevitably, the major concern is medicine; however a relatively substantial treatment (5 pages) of the virology literature can be found under a section entitled 'Medical microbiology'. This is an account, largely from a microbiological viewpoint, of dictionaries, handbooks, general texts, author collections and reference works. In addition 80 books, serials and bibliographies are listed, with some critical assessment. The author, who is a university lecturer in microbiology, writes clearly with the user very much in mind.

From the same series as the aforementioned guide comes *The use of biological literature* (Bottle and Wyatt). However, it is nowhere near as useful as its stable companion, partly because of its age (1971) and partly because of its rather superficial coverage of virology. Virology gets little more than a passing reference, with 50 or so books and a few serials mentioned briefly. Of interest also are the sections on 'Tissue culture' and 'Immunology', the chapter on 'History and biography of biology', and the appendices containing lists of British libraries and culture collections. A new edition is being planned for 1984.

Guide to the literature of the life sciences (Smith and Reid, 1972) adopts a somewhat different approach as compared with the other guides, being much more discursive in nature, dealing with information problems rather than information sources. A limited number of virology textbooks, periodicals and taxonomic texts are furnished as examples.

Information sources in agriculture and food science (Lilley, 1981) is rather disappointing in its minute coverage of virology, which is found in the chapter on 'Veterinary science'. However it is useful in that it furnishes a very current account (to 1980) of a field into which virologists often make incursions. The chapter on computer-based bibliographic services in agriculture may also be of interest both to plant and animal virologists.

Closer to home in terms of subject coverage is the National Cancer Institute's publication *Directory of cancer research information resources* (2nd edn, 1979), which provides cancer researchers with a single-volume listing of most of the available cancer information services internationally. Many virological and virological-related primary and secondary journals with a cancer bent are listed. It is arranged into 12 sections: 'Primary publications', 'Secondary publications', 'Classification schemes', 'Libraries', 'Special collections', 'Computer-based information systems and services', 'Audio-visual information services', 'Dial-access services', 'Cancer registers', 'Research projects information sources', 'Organizations US' and 'Others', with four indexes: title, geographic, organization and subject. Because of the close link between viruses and cancer this directory is a very useful source, its use being that much greater because of its relative currency.

5.2 TRACING AND LOCATING JOURNALS (*see* bibliography entries 305–314)

There are no guides to journals published at the virology level; however, there are some useful ones published on a more general level. Whilst one of our intentions in providing the detailed and comprehensive account of virology journals found in section 4.1 was to dispense with the need for these, nevertheless because of the necessity of keeping up to date with the latest changes in the journal literature, and the desirability of obtaining a somewhat broader view of it, the most important of them merit discussion.

One of the most comprehensive listings of virology journals can be found in a bibliography published by the International Association of Microbiological Societies, and called *Periodicals relevant to microbiology and immunology*. Although covering 700 periodicals, from 35 countries, that regularly or sporadically publish material of microbiological or immunological interest, its value is much diminished by the fact that it last appeared in 1968. Since that date of course much has changed in the journal literature: many new journals have appeared and many of the existing ones have since changed frequency, format, sponsor, etc. The list is, then, chiefly of historical interest.

Today for the most current and comprehensive coverage of journals we must look to *Ulrich's International periodical directory*. It is an annual classified listing, equipped with a title index, which provides all the details necessary to enable a periodical to be traced or bought. An attempt is also made in a rather abbreviated way to provide some indication of a journal's contents (i.e. whether it contains reviews, illustrations, statistics, bibliographies). Unfortunately virology journals do not merit their own place in the classification so it is necessary to browse through the lists of microbiology and medical journals, amongst which they are placed. All told there are 11 virology journals listed, with only a few major omissions (e.g. *Virusy i Virusnye Zabolevaniya*), but perhaps *Ulrich's* greatest weakness is that the only way it can be used to identify virology journals is from words in their title. (In theory one should be able to distinguish them by the Dewey Decimal classification number — in virology's case 576.64 — that accompanies each entry but numbers are allocated rather broadly, most receiving the microbiology number 576.) There is no way by which those journals of other disciplines that regularly publish articles on viruses can be identified; and as we have seen that many of the most important papers in the field appear in these related journals, then *Ulrich* and bibliographies like it are necessarily of limited value. A companion volume to *Ulrich* is *Irregular serials and annuals*. Identical in format, this biennial publication lists annuals, yearbooks, reviews and reports issued irregularly or less frequently than once a year. The two companion volumes may be searched on-line via either the DIALOG or BRS networks.

To keep abreast of changes in the journal literature in between annual volumes it is necessary to consult *Ulrich's Quarterly*, which keeps exactly to the same format as its parent. The *British National Bibliography (BNB)* can be used in this way too for it lists new periodical titles on a weekly basis. Its subject format makes it possible to go straight to the virology section to see whether any new titles have appeared. *New Serial Titles*, of which more will be said later, undertakes a similar function for US periodical titles, though monthly rather than weekly.

A possibly more unconventional source of information on virology journals — but one against which the previously voiced criticism of being led only to journals entirely devoted to virology cannot be levelled — is *Virology Abstracts*. Once a year the publishers produce a separate list of the journals they scan for all their abstracting services, not just *Virology Abstracts*; these of course include the journals of many disciplines. Ironically the very interdisciplinarity of the list creates its own problems because one is confronted with literally pages of journals (5000 approx.), many of which are only occasional or infrequent providers of virological articles. The list is also circumscribed in another way: the journals are not adequately described, such descriptions not being seen as its prime purpose.

If it is a question of borrowing a particular issue of a periodical or obtaining a photocopy of a periodical article then the appropriate tool is *Current serials received*, which lists alphabetically by title the 51 500 serials the British Library's Lending

Division receives. Only sufficient details to enable the journal's tracing (title and shelf location mark) are provided with each entry. For those simply seeking the titles of virology journals, for which no names are known, then another British Library Lending Division (BLLD) publication is the answer. *Keyword Index to Serial Titles* is an alphabetical list of all significant words in the titles of journals. Under each word all the titles containing that word are listed. Thus each journal may be listed in a number of different places, boosting its chance of retrieval. The adoption of such an approach also means that one can still find a journal even without knowing the precise title. The *Keyword Index* is in fact an index not just to the 51 500 current journals held by the BLLD but to its total holdings (140 000) of live and dead serials. The index is held on about 65 fiche and is updated every quarter.

Another British reference work that shows the whereabouts of periodicals is the *British Union Catalogue of Periodicals (BUCOP)*. It was cumulated annually into two volumes, one of which is called the *World list of scientific periodicals*, the other listing all serials regardless of subject. The *Catalogue* provides a guide to the serial holdings of British libraries, including the British Library, and is invaluable to anyone wanting to browse through journal back numbers rather than borrow specific articles, or consult long-dead serial titles. However its usefulness to the virologist is limited by a number of its features. First, it affords an approach only by title (you have to know the name of the journal you are looking for). The fact that it has an index to the journals' sponsoring bodies is a mitigating factor, because for anyone armed with a knowledge of the names of virology organizations it is just possible to keep a weather eye out for new publications in the field. Secondly, the lack of one major cumulated sequence (there are four cumulations, and 10 annual volumes) means that a large number of different volumes have to be scanned when seeking particular journals. Thirdly, it is grossly out of date in its recording of libraries holdings and with its recent demise (1980) the position has obviously worsened considerably.

Its place has largely been taken by *Serials in the British Library*, a quarterly list that affords much more detail for each journal than *BUCOP* supplied. It is arranged alphabetically by title and lists only newly acquired serials: those acquired by the British Library and a few specialist libraries. It cumulates annually on microfiche. (The fiche actually contains titles not listed in quarterly issues.) Its value is of course circumscribed by its lack of a subject index.

The rough equivalent of *BUCOP* in the USA — *New Serial Titles* — is far more helpful. Like *BUCOP* it lists the serial holdings of US libraries, including the Library of Congress, but that is as far as the similarity goes. Most importantly of all, *New Serial Titles* is available in both alphabetical title and subject forms. The latter is of prime interest to us as it enables us to browse for virology titles. The bibliography is arranged according to the Dewey scheme (*see* section 5.8), which means that relevant titles can be found amongst microbiological serials at 576 and medical ones at 616.9 — the disease number. Invaluable also is the fact that *New Serial Titles — Classed Subject Arrangements* appears monthly, thus providing up-to-date intelligence of periodical holdings. However the problems of cumulation beset *New Serial Titles* too. The major cumulation covers the period 1950–70. The title-arranged edition exists also in fiche form for this period. A 1971–5 cumulation is available in the title format but not in subject format.

A word ought to be said here about one of the largest bibliographic computerized databases in the world: OCLC. Though centred in Ohio, OCLC can be accessed from most countries in Europe. Around a quarter of a million serial titles can be traced to a location through the OCLC system (inter-loan facilities are available to participating libraries). A search can only be conducted via the title of the journal, although it is intended to provide access by subject, in the not too distant future.

5.3 TRACING AND LOCATING PERIODICAL ARTICLES

There are a rich variety of services whose almost sole aim is to list and report on the publication of journal articles. Something like 90% of all recorded information in science is disseminated in article form so it is perhaps not so surprising that so many bibliographic services owe their existence to just one communication form. First and most important are the abstracting and indexing services; second the relatively new current-awareness services; third the reviewing services, which are largely selective in their listing (because virologists tend to read these as they would any other scientific journal — for subject information rather than bibliographic references — they are treated separately in section 4.2); and finally the subject bibliographies which, unlike the others, are normally 'one-off' publications.

Abstracting and Indexing Services

Whilst it is perfectly possible to judge a book by looking at its cover the same certainly cannot be said of journals. The journal itself, in science anyway, is really only a means for transporting information, which is largely packaged in the form of articles. It is mainly because these articles lie buried somewhat anonymously between the journal's covers that abstracting and indexing journals have arisen: they provide the invaluable service of making an article's publication much more visible. They do this by: providing a variety of search pathways to articles and papers (author, title, subject); drawing together articles published in different journals but on the same topic; and in the case of abstracting journals, summarizing their contents.

Abstracting and indexing journals are current-awareness tools in the sense that they alert users to new articles that might be of interest. Because they must adopt some form of classified arrangement they serve as a broad guide to new developments in the field. Many see it as a particularly important function to trawl widely in pursuit of articles unlikely otherwise to be seen by their readership. *Virology Abstracts* for instance sees itself very much as a solution to the problem of scatter, which is so endemic in interdisciplinary fields such as virology; that is, it provides a structure and coherence that is otherwise missing from the literature.

In science some abstracting/indexing services have become so good at their job that they virtually offer the prospect of locating any and every article that has been published in the field since the service's inception. Thus *Virology Abstracts* offers an archive constituting 20 years of the virology literature — effectively something like 90% of the whole archive. Unfortunately the end-product of such diligence is vast numbers of references, many of dubious worth, which, ironically, in themselves constitute an information problem too big for most to cope with.

Another problem encountered is likely to be the speed with which abstracting and indexing services report on an article's publication: there is inevitably a time lag of some months. In the case of *Virology Abstracts* this varies between 4 and 6 months. Now this gap in listing can be quite crucial, for by definition the most current data are likely to be most useful or, at least, the least likely to have been encountered. To fill this gap publishers have resorted to current-contents lists and more rudimentary forms of indexing (i.e. rotating the words in a journal article's title).

With all large and complex abstracting and indexing journals it is easier, quicker and more efficient to search them via their computerized equivalents. This is especially the case when a retrospective search is being conducted. If one is looking for a needle in a haystack — and one often has that feeling when searching through a service that is generating over 100 000 references per year (about par for abstracting journals in science) — it is obviously better to do it with the aid of a computer.

Although virology has its very own abstracting service in *Virology Abstracts*, which most will find more than meets their needs for intelligence on newly published articles, it is unfortunately still necessary to consider some of the abstracting and indexing services of other disciplines. There are three reasons for this: first, because of the way in which the virology literature is scattered throughout the journals of many fields, a totally comprehensive view of virology publishing can be obtained only by scanning the abstracting and indexing journals of these fields also; secondly, each abstracting/indexing service analyses and describes an article from the point of view of its own particular readership (e.g. for the veterinary implications of a piece of virus research one would normally obtain a description from, say, *Veterinary Bulletin*); thirdly, if it is a literature search back in time that is to be conducted, then *Virology Abstracts* can offer only a period stretching back to 1967; thus a service such as *Biological Abstracts*, which started in 1927, would also have to be consulted. (It would probably not be wise to trust *Virology Abstracts*' coverage even back to 1967, for in its formative years, say to the early 1970s, coverage would almost certainly have been selective.)

Ten abstracting and indexing services are judged to have some importance to virologists. These are listed in *Table 5.1*. Of these, five — *Biological Abstracts, Excerpta Medica, Index Medicus, Index Veterinarius* and *Virology Abstracts* — are of central concern and have been treated in somewhat more detail; moreover these titles have been subjected to a special test to ascertain their relative performances in respect to coverage and speed of reporting. *Science Citation Index* has also been singled out for detailed comment, not so much because of its central importance but because of the unique approach it adopts.

Of historical interest only is an indexing journal called *Virology Literature*, edited by R. S. and M. D. Schaiffenburg, which was launched in 1973 but published only three issues before it ceased.

Individual Abstracting and Indexing Services

(*see* bibliography entries 315–329)
Biological Abstracts

Biological Abstracts is published twice a month by Biosciences Information Service and is one of the largest and most comprehensive abstract services of all, containing

Table 5.1 Main indexing and abstracting journals in virology

Journal	Field covered	Began publi-cation	Frequency	No. of journals scanned	Approx. no. of total abstracts p.a.	Approx. no. of abstracts on virology p.a.
Biological Abstracts	Biology	1927	Semi-monthly	9 000	165 000	1 700
Excerpta Medica, section 47	Virology	1971	Approx. monthly	4 000	3 800	3 800
Index Medicus	Medical and animal virology	1879	Monthly	3 000	500 000	5 000
Index Veterinarius	Veterinary	1933	Monthly	1 350	23 000	1 000
International Abstracts of Biological Sciences	Biology (much phage)	1954	Approx. monthly	500	35 000	2 400
Review of Plant Pathology	Plant	1922	Monthly	1 400	6 300	1 000
Science Citation Index	Science	1961	Bi-monthly	3 000	500 000* 2 000 000†	10 000 (est.)†
Veterinary Bulletin	Veterinary	1931	Monthly	1 350	8 000	1 000
Virology Abstracts	Virology	1967	Monthly	5 000	9 000	9 000
Zentralblatt für Bakteriologie, Parasitenkunde, Infektionskrankheiten und Hygiene	Microbiology	1879	Monthly	400	4 300	1 200

* References; † Citations

over 150 000 abstracts a year and scanning 9000 primary journals as well as government reports, symposia and reviews. Together with *Biological Abstracts/ RRM* it constitutes the major English-language service providing coverage of the life sciences. *Biological Abstracts* is also available in machine-readable form. On DIALOG it is called *BIOSIS Previews*. Well over three million publications issued since 1969 are listed on two files. (It is also available on ORBIT.)

Abstracts are listed under the following subject headings:

    Biochemistry
    Genetics of bacteria and viruses
    Medical and clinical microbiology (includes Veterinary)
    Microorganisms, general
    Neoplasm and neoplastic agents
    Phytopathology
    Tissue culture, applied methods and media
    Veterinary science
    Virology, general
        animal host viruses
        antiviral agents
        bacteriophages
        general methods
        genetics
        immunology
        pathogenic, medical
        pathogenic, viruses
        plant host viruses
        viruses

Every volume has a subject guide that lists all the synonyms that refer to related fields of interest. Each issue has a table of contents listing the various indexes:

    author — used to find content summaries by personal or corporate names listed A–Z
    biosystematic — used to find content summaries by taxonomic categories
    generic — used to find content summaries according to genus–species names
    concept — used to find content summaries relating to broad subject areas of biology
    subject — used to find content summaries from specific words appearing in the author title or index

A typical abstract is set out like this:

VIROLOGY GENERAL[1]
Animal host viruses[2]

10788 Weinburg, Robert A [address] CELL **22** (3) 643–644 (1980)
Origin and roles of endogenous retroviruses/REVIEW, ONCORNAVIRUS, CHICKEN, MOUSE, FELIDAE, MONKEY, VERTEBRATE, COW, HUMAN, TUMORIGENESIS, EVOLUTION, GERMLINE, PROVIRUS[3]

CON[4] Animal cytology and cytochemistry/Animal genetics and cytogenetics/ Carcinogens and carcinogenesis/Evolution

TAX[5] Retroviridae; Oncornoviridae/Galliformes/Hominidae/Bovidae/Primates/Muridae/Felidae

---

[1] Major heading   [2] Sub-heading   [3] Subject terms
[4] Subject concepts   [5] Taxonomic names

To supplement *Biological Abstracts*, a computer-produced monthly publication prepared by Bioscience Information Service (BIOSIS) and entitled *Biological Abstracts RRM* (formerly *Bioresearch Index*) was started in 1967. It provides access to more than 100 000 research papers annually in addition to those reported in *Biological Abstracts*. Only citations are given; no abstracts. Each issue is composed of six parts: a permutated title index, giving a subject approach through keywords; a bibliographic section, arranged by journal title and issue; an author index; the cross index; the subject index (BASIC); and the biosystematic index. Also available on DIALOG.

Excerpta Medica

In all there are 44 separate sections of this monthly abstracting service and since 1971 there has been one devoted to virology — section number 47. Prior to this, virology had to share space in *Microbiology — Bacteriology, Virology, Mycology and Parasitology*, which began in 1948.

The classification scheme used in the virology section is as follows:

1. General Aspects
2. Classification, Morphology and Replication
3. Epidemiology, Infection, Pathogenesis and Pathology
4. Physiology
5. Biochemistry and Genetics
6. Immunology
7. Disease Prevention
8. Interferon and Interference
9. Therapy, Chemotherapy
10. Techniques

11. Isolation and Cultivation
12. Inactivation, Conservation
19. RNA Viruses
20. RNA Viruses (up to Dec. 1978)
21. DNA Viruses
22. Unclassified Viruses, Viroids and Other Agents
30. Bacteriophages
31. Plant Viruses

In 1980 nearly 3800 twenty- to thirty-word virological abstracts appeared. Author and subject indexes are contained in each issue and these are cumulated annually in the last of 10 or 12 issues which comprise each volume.

A typical abstract is

19.6   Rhabdoviridae
19.6.2   Lyssavirus
3548   Population dynamics of fox rabies in Europe. Anderson R. M., Jackson H. C., May R. M., Smith A. M. [Address............]
         NATURE (London) 1981 289/5800 (765–771) [Abstract............]

This will be found in the subject index as:

Rabies — disease, transmission, fox, population dynamics, mathematical model   3548

and in the author index under:

Anderson R. M.   3548

In addition to section 47, which is exclusively virological, a small amount of information on viral-related subjects can be found in other sections as well:

Antiviral drugs	30 Pharmacology and Toxicology
	37 Drug Literature Index
Biochemistry	4 Microbiology
	29 Biochemistry
Cancerogenic Viruses	5 Pathology
	16 Cancer
Epidemiology	17 Public Health
Genetics	22 Genetics
Immunology	26 Immunology, Serology and Transplantation
Viral Hepatitis	48 Gastroenterology

Excerpta Medica is also available on-line via DIALOG. About two and a half million abstracts of documents published since 1974 are available in this way. Unusually quite a few of these abstracts do not appear in the hard-copy equivalent. Another useful feature is that items being processed are also noted; this can help to ameliorate the time lag problem.

ICRDB Cancergrams

The International Cancer Research Data Bank (ICRDB) of the US National Cancer Institute compiles and publishes monthly bulletins called *ICRDB Cancergrams*, which contain abstracts of recently published articles on 65 specific cancer topics. Four of these deal specifically with viruses:

RNA Viruses Associated with Cancer—Cell Biology and Animal Studies
RNA Viruses Associated with Cancer—Molecular Biology
Viruses of Humans and Other Primates
Viral Immunology

Index Medicus

Published under various titles by the National Library of Medicine since 1879, this complex indexing journal consists of a monthly listing of bibliographic references to current articles. Entries are arranged under narrow subject headings, and included in each issue is an author index. A separate section entitled 'Bibliography of medical reviews' accompanies each issue.

The individual parts of *Index Medicus* are superseded by a cumulated annual issue known as *Cumulated Index Medicus*, which for the year 1980 amounted to 14 volumes containing approximately 245 000 citations.

Published as Part 2 of the January *Index Medicus* is 'Medical subject headings' (MeSH), which contains the subject headings used in the journal, arranged alphabetically with cross-references and in categorized lists. The annually revised subject headings, under which all citations appear, serve as the basis for search formulations in retrieval of bibliographic citations stored in the equivalent computerized information retrieval service, known as MEDLINE.

The categories and subcategories relating to virology are:

- B   Organisms
  - B4   Viruses [over 300 listed A–Z mainly in families and in their modern forms]
- C   Diseases
  - C2   Virus diseases [over 150 listed A–Z]
  - C4   Neoplasm [some of viral aetiology]
    e.g. C4 557.386.546 Marek's disease
  - C10   Diseases of nervous system (*see also* other organs)
  - C10   228.228.674.778 Poliomyelitis
- D   Chemicals and Drugs
  - D20   Anti-infective agents
  - D20   388.643 Interferon
- G   Biological Sciences
  - G1   Biological sciences
  - G1   273.540 Microbiology
  - G1   273.540.859 Virology

A typical entry in the monthly listing of references would be:

MAREK'S DISEASE VIRUS
B4.909.204.382.675   (Herpesviridae)
B4.909.574.204.525   (Oncogenic DNA viruses)

75 [citations]

x   Fowl Paralysis Virus [synonyms]
x   Neurolymphomatosis Virus

Such a massive undertaking as *Index Medicus* obviously gives rise to difficulties in handling and inevitably causes a great scattering of information relating to virology, as can be seen from the above examples. It should also be noted that contrary to the medical connotations of the title of this journal, viruses and virus diseases of animals, such as Marek's disease virus, are also covered.

It is probably preferable to use the on-line form of this bibliographic tool — MEDLINE — because (a) it is available for searching before the printed *Index*; (b) it provides a greater number of access points (12 as opposed to 3); (c) material is cumulated automatically, providing one-place reference to large parts of the file (on DIALOG the database is held on three files: 1966–72, 1973–79 and 1980 onwards). On DIALOG (it is also available on BRS and BLAISE) the MEDLINE database constitutes around 4 000 000 records drawn from the journals of more than 70 countries. Over 40% of the records added since 1975 contain author abstracts.

Index Veterinarius

The Commonwealth Agricultural Bureaux (CAB) compile and publish the veterinary equivalent of *Index Medicus*. *Index Veterinarius* regularly scans 1250 serial publications, books, annual reports, monographs, theses and other non-serials. Each issue arranges publications both by subject and author with each entry listed under one or more subject headings in the subject part, and also in the author part. Subject headings are drawn from the controlled vocabulary currently in use at the CAB, copies of which can be obtained from them. This was last revised in 1979.

General virological material will be found under the subject headings:

Viral Diseases
Viral Immunosuppression
Virology

The main bulk of the information will, however, be found under the name of the virus, the disease it causes and/or the animal it infects. For example, Finnie, J. Canine parvovirus infection. *Victoria Veterinary Proceedings* **37**, 12–13 (1979). This paper will be found in the subject part under 'Dog diseases', 'Enteritis', 'Parvoviridae' and 'Mycocarditis' and under Finnie, J. in the author part. Not all papers are, however, so comprehensively categorized. For example, a paper cited from *Intervirology* by Campbell, W. F. *et al.* **10** (1), 11–23 (1978) on Marek's disease virus was found under the heading 'Marek's diseases' but was not found under 'Poultry diseases'.

Some of the papers indexed have been abstracted in *Veterinary Bulletin* and where this is the case a reference is made.

Index Veterinarius together with *Review of Plant Pathology* and *Veterinary Bulletin* may be searched simultaneously on the Commonwealth Agricultural Bureaux' computerized database, CAB ABSTRACTS. CAB ABSTRACTS is offered on-line by DIALOG amongst others. The service, which covers items published since 1973, is updated monthly. Only 'significant' papers obtain abstracts; others are limited simply to the standard bibliographic details.

International Abstracts of Biological Sciences

Previously known as *British Abstracts of Medical Science, British Abstracts* and *British Chemical Abstracts, International Abstracts of Biological Sciences* (*IABS*) is published by Pergamon Press and covers those aspects of experimental biology which it considers are not covered by the CAB's and BMA's abstracting services.

It contains abstracts and titles or expanded titles obtained from 100 'core' journals and titles or expanded titles from another 400 journals. Selected titles of symposia, etc. are published in a separate section. Review articles appear in the relevant journal sections. There is an author index.

IABS is divided into 12 disciplines, of which two are of interest to virologists: 'Microbiology' and 'Genetics'. These themselves are further divided:

MICROBIOLOGY
  Methods and general studies
  Identification
  Isolation and culture
  Development and structure
  Composition and metabolism
  Pathogenesis and epidemiology
  Ecology
  Inhibition
GENETICS
  General
  Bacteria and viruses
  Other microorganisms
  Plant
  Animal
  Human
  Effects of radiation and chemical agents

Because of the layout of the scheme, with its bias towards bacteriology, the virology abstracts fit uneasily amongst subdivisions that are neither helpful nor relevant to the virologist.

Review of Plant Pathology

This monthly journal, prepared by the Commonwealth Mycological Institute, contains abstracts pertaining to infections of plants. It has an author index and a subject index. The abstracts are largely classified alphabetically by crops.

The contents of interest to plant virologists are headed:

Reports, annual conferences and symposia
Viruses
Techniques

In addition, under each crop, abstracts are arranged in the order:

General
Virus and mycoplasma diseases
Fungus diseases
Bacterial and physiological

A typical entry would be: 'Properties of tobacco yellow dwarf and bean summer death viruses'. Thomas J. E. and Bowyer J. E. *Phytopathology* (1980) **70** (3) 214–217. In the subject index the paper will be found under:

'Bean'
Summer death virus, in Australia
properties and transmission

and

'Tobacco'
Yellow dwarf virus, in Australia
properties and transmission

The reference will also be found under the name of the author in the author index.

Science Citation Index

This interdisciplinary index to the scientific literature offers an approach somewhat different from the others, although the approach is not as novel as it first might seem for it capitalizes upon the time-honoured method scientists have used to follow up their interest in a topic; that is, by examining the citations or references at the bottom of an article of interest. Indeed surveys have shown this is by far the most popular way of searching the literature, the idea being of course that if the article is relevant then so too would be the articles the author cites in support of his or her paper.

Thus what *Science Citation Index (SCI)* does is simply to gather together all the citations appended to articles, letters, editorials or chapters in the journals and books it covers. It covers annually over 5000 documents of which about 3300 are journals. All the main virology journals are in fact covered. More importantly, because it is an interdisciplinary publication — its boundaries being prescribed only by what scientists cite — it will pick up virology articles, etc. whatever the discipline of the journal in which they are published, an extremely important point since much of virology's literature is published outside its own boundaries, and precisely where outside is certainly not always predictable.

At the very centre of *SCI* (a three-part work) lies the citation index where, as mentioned above, all citations are gathered together in one alphabetic sequence by the cited author. Under each author is listed the particular works of his/hers that are cited. Appended to each cited publication are the authors (called source authors) who have cited it. The value of such an arrangement is that if you know that a particular paper is central to your work then you can find related papers by finding out who has cited that paper (the assumption being that the interests of the two papers are similar). Such a method can relate papers much more effectively then the subject classifications adopted by the more conventional abstracting and indexing services because it is the authors (i.e. the subject experts) who are making the subject links. Furthermore, in theory anyway, one supposes that citation indicates value and thus only the best, most worthy, unique, etc. find themselves indexed by *SCI*. Thus citation indexing provides in theory a degree of selectivity not normally associated with abstracting and indexing services. (The sheer bulk of the publication of course appears to argue against this.)

The *Source Index* and the separately published subject approach to it (*Permuterm Subject Index*) provide details of the items from which the citations forming the *Citation Index* were taken. They themselves form additionally a current indexing service and can be used separately in the normal way that we use the other abstracting and indexing services. The *Source Index* in reality comprises three indexes, the first a corporate index listing currently published articles by the geographical location and name of the institutions to which the authors of these articles belong. Thus from this index it is quite possible to establish the productivity of the various research establishments in virology. If the geographical location of the establishment is not known then the second index should be consulted, which lists organizations alphabetically, each organization being accompanied by the details necessary to enable it to be located in the geographic index, where full details of publications are provided. Thirdly, there is the source index proper: an alphabetic listing of all the authors of papers from which citations were extracted. Here full bibliographic details are provided to enable their location. The very abbreviated style of the reference does take some getting used to, as does the dense format adopted by the whole work. In this respect it is certainly not 'user friendly' and the reader may be advised to access it in its on-line form (it is available on ORBIT and DIALOG).

The *Permuterm Subject Index* provides the necessary subject pathways to the literature. It is an alphabetic listing of pairs of significant words taken from the title of publications listed in the *Source Index*. The success of the subject search is inhibited by the vagaries of the natural language and additionally by the fact that virology documents can be found under dozens of subject headings, *virus, virology, virological* being the most important. These three headings alone yield somewhere near 1900 entries each bi-monthly issue, making *SCI* an important source for virologists despite the drawbacks of its layout and bulk.

SCI is known on DIALOG as *Social Sciresearch*. In machine-readable form it is available back to 1972 and constitutes a file of approaching one and a quarter million records.

Perhaps the greatest use made of *SCI* is a rather vain one: researchers scan it to see whether they have been cited and by whom.

Veterinary Bulletin

Companion journal to *Index Veterinarius*, the *Bulletin* is published monthly by the Commonwealth Agricultural Bureaux and contains abstracts of veterinary interest, totalling some 8000 a year. Each issue also carries a review article, many of which have been virological in nature, an example being 'Canine adenoviruses' (1981). The reviews, incidentally, are available as reprints.

The abstracts themselves are divided into broad subject categories with papers concerning virology found under the headings:

Neoplasms
Virology
Viral diseases

Under these headings the subject is haphazardly classified into the various fields of interest.

For example, under the heading 'Viral diseases' can be found the following: 'Foot-and-mouth'; 'Togavirus infections'; 'Viruses of horses'; and 'Viruses of small animals'. Under 'Neoplasms and leukoses' will be found 'Oncogenic viruses', such as Marek's disease and feline leukaemia virus.

Virology Abstracts

Information Retrieval Ltd began publication of *Virology Abstracts*, an important abstracting service for virologists, in 1967. Each monthly issue contains approximately 750 abstracts followed by an author index and subject index, which are cumulated annually. The subject index is compiled from chains of keywords, which are selected from a fixed term list. Several chains are assigned to each abstract, and are rotated to provide entry points for significant keywords.

The staff monitor approximately 5000 primary journals and other reference sources for papers falling within the scope of *Virology Abstracts*. Abstracts are usually 150–200 words in length and outline the content of the paper, the methods used, the results obtained and conclusion drawn. Abstracts of reviews and monographs are usually shorter and describe briefly the main points covered.

Entries are divided into the following categories:

Viral taxonomy and classification
Methodology and tissue culture studies
Physicochemical properties, structure and morphology
Replication cycle
Viral genetics
Phage–host interactions
Immunology
Antiviral agents
Oncology
Viral infections of man

Diseases associated with slow viruses
Viral infections of animals
Animal models
Viral infections of invertebrates
Viral infections of fungi and lower plants
Viral infections of higher plants
Miscellaneous topics

These main categories may be subdivided. A typical example is:

ANIMAL MODELS AND EXPERIMENTALLY INDUCED
VIRAL INFECTION OF ANIMALS

5855–V14   Experimental infection of broiler chickens with an avian reovirus. Wood G. W., Thornton D. H.   *J. Comp. Pathol.* **91** (1) 69–76 1981.

This abstract will be found in the subject index under:

> Avian tenosynovitis virus
>   experimental infection, chickens   5855

and

> Chickens
>   avian tenosynovitis virus   5855

Curiously the abstract is not found under 'Avian reovirus' or 'Reovirus'.
  The abstract number will also be found in the author index under Wood G. W. and Thornton D. H.
  Virology Abstracts is held on-line, together with another 14 printed series of abstracting journals (the most relevant of which are *Biochemistry Abstracts*, *Genetics Abstracts*, *Immunology Abstracts*, *Microbiology Abstracts* and *Oncology Abstracts*) on a file called *IRL Life Sciences Collection*. Access to the file can be obtained via the DIALOG network. The file goes back only to 1978, so offers limited retrospective searching, but even so it exceeds a quarter of a million records. The *Collection* is updated monthly and as with other on-line services offers the prospect of conducting a search based on any word in the title, abstract or accompanying index string. The great value of being able to search all these abstracting services at the same time needs little stressing.

Zentralblatt für Bakteriologie, Parasitenkunde, Infektionskrankheiten und Hygiene

Despite its title, this journal, with the exception of a few book reviews, is published in English. In fact it is subtitled *Abstracts of Microbiology and Hygiene* and is published by Gustav Fischer Verlag. Each issue carries about 200 abstracts on microbiology. It is broadly divided into:

    General medical microbiology
    Methods of microbiology
    Micrococcae and other bacteria
    Virology
    Book reviews (in German)

Virology is further divided:

    General
    Major virus families (named)
    Slow virus infections
    Tumour viruses
    Other medically interesting viruses
    Unclassified viruses
    Phages
    Insect and plant viruses

Within the virus families the abstracts appear to be in no particular order.

Other Abstracting and Indexing Journals of Interest to Virologists

    Bulletin de l'Institut Pasteur
    Bulletin Signalétique
    Chemical Abstracts
    Microbiologica Synthesis
    Tissue Culture Abstracts
    Tropical Diseases Bulletin and *Abstracts on Hygiene*

Bibliographic details of these sources are not included.

Comparative Assessment of the Abstracting and Indexing Journals Most Commonly Used by Virologists

In order to examine the relative strengths of the most widely used bibliographic periodicals, we took 12 original papers from the *Journal of General Virology*, the *Journal of Virology* and *Virology*. Selection of the papers was largely at random but represented, equally, the four main branches of virology: animal, human, bacterial and plant virology. Attempts were then made manually to trace these articles using the hard-copy editions of virology's main abstracting and indexing journals:

    Biological Abstracts
    Excerpta Medica

Table 5.2 Comparative assessment of the most commonly used abstracting and indexing journals in virology

Journal article sample	Subject area	Journal				
		Biological Abstracts	Excerpta Medica	Index Medicus	Index Veterinarius	Virology Abstracts
'Structural polypeptides of mumps virus' C. Orvell. *J. Gen. Virol.* Dec. 1978.	Human	Yes	Yes	Yes	No	Yes
'Early events in the infection of tobacco with alfalfa virus.' L. Van Vlotendoting. *J. Gen. Virol.* Dec. 1978.	Plant	Yes	No	No	No	Yes
'The scrapie agent: evidence against its dependence for replication on intrinsic nucleic acid.' T. Alper et al. *J. Gen. Virol.* Dec. 1978.	Animal	Yes	Yes	Yes	Yes	Yes
'Studies on the transduction process by SPP1 phage.' E. Ferrari et al. *J. Gen. Virol.* Dec. 1978.	Bacteria	Yes	No	Yes	No	Yes
'Genomic stability of gibbon oncornavirus.' L. Sun et al. *J. Virol.* Dec. 1978.	Animal	Yes	No	Yes	No	Yes
'Control of replication in RNA bacteriophages.' P. Pumpen et al. *J. Virol.* Dec. 1978.	Bacteria	Yes	No	Yes	No	Yes

Title						
'No homology between double-stranded RNA and nuclear DNA of yeast.' N.D. Hastie et al. J. Virol. Dec. 1978.	Plant	Yes	No	Yes	No	No
'Specific secretion of polypeptides from cells infected with vaccinia virus.' M. A. McCraw et al. J. Virol. Dec. 1978.	Human	Yes	Yes	Yes	No	Yes
'The role of the bacteriophage λ F1 gene product during phage head assembly in vitro.' S. Benchimol et al. Virology. Dec. 1978.	Bacteria	Yes	Yes	Yes	No	Yes
'Phosphoproteins of the murine mammary tumor virus.' N. H. Sarkar. Virology. Dec. 1978.	Animal	Yes	Yes	Yes	No	Yes
'Stabilization and particle morphology of prune dwarf virus.' E. K. Halk et al. Virology. Dec. 1978.	Plant	Yes	No	Yes	No	Yes
'Purification and characterization of proteins excreted by cells infected with herpes simplex virus, etc.' A. B. Chen et al. Virology. Dec. 1978.	Human	Yes	Yes	Yes	No	Yes
Total		12/12	6/12	11/12	1/12	11/12

Index Medicus
Index Veterinarius
Virology Abstracts

Results of this exercise are summarized in *Table 5.2*. It can be seen that *Biological Abstracts* listed all 12 papers, and *Virology Abstracts* had all but one, the omission being a paper on viral-like nucleic acid in yeasts. *Index Medicus* similarly missed one — a paper on plant viruses — although, rather surprisingly, it contained the other two, also dealing with plant virology. *Excerpta Medica* contained only half of the 'test' papers; included were all of those dealing with human viruses, all but one of those on animal viruses, one of the phage papers but none of the plant virus papers. *Index Veterinarius* performed poorly, with only one of the 'test' papers being covered, this (typically) being one devoted to a virus of sheep.

Whilst comprehensiveness is important in an abstracting service, how quickly the abstract arrives on the shelf is equally crucial. It should be noted here that the printed date on the abstracting journal does not always coincide with its actual date of arrival at the library. This may be several months later than the advertised date.

Using the later date in our calculations, on average *Biological Abstracts* and *Virology Abstracts* proved most current with a time lag of six months. The quickest they managed to abstract an article was five months and the longest eight months. *Index Medicus* showed greater variation around the mean of seven months: one paper appeared within four months and another after as long as eleven months. Because we had only one article from *Index Veterinarius* upon which to base our calculations, a number of other papers were examined for their lateness in appearance; this averaged between five and six months.

In this study, searching for the papers was facilitated by having the authors' names but in a real situation the relevant sections or keywords must be laboriously scanned. As said previously, the classification schemes used by the different journals differ considerably, thus the location of the papers will also differ. For example, 'The scrapie agent: evidence against its dependence for replication on intrinsic nucleic acid' by Alper *et al.* will be found under 'Viroids' in *Excerpta Medica*; under 'Replication cycle' in *Virology Abstracts*; and under 'Scrapie' in *Biological Abstracts*, *Index Veterinarius* and *Index Medicus*.

Current-Awareness Publications (*see* bibliography entries 330–332)

Keeping up to date with what is being currently published is no easy task, with hundreds if not thousands of journals all being sources of potentially relevant information. Most people, if not because of the sheer time involved but because of the difficulties in finding somewhere that takes all the journals, resort to reference tools that are specifically designed to scan and monitor the literature on their behalf. The conventional abstracting and indexing journals were initially founded to fulfil just such a role but because they cannot process the data sufficiently quickly (six months is common) for some scientists (although it must be said that from the limited opinion samples we obtained from virologists most would seem content with the performance of the standard abstracting and indexing journals) a new breed of publication, the contents listing, has been introduced to trim back some of this delay.

Current Contents, published by the Institute for Scientific Information, is a series of just such publications. The two titles most relevant to virologists are *Current Contents — Life Sciences* and *Current Contents — Agriculture, Biology and Environmental Sciences*. Both services are published weekly (if any claims of currency are to be made such a frequency is virtually essential) and both cover more than a thousand journals, although this total is not reached each week because of the variation in the publication patterns of individual journals. At the centre of these two publications are the facsimile copies of the contents pages of the journals covered in that particular issue. In general these contents pages are only two weeks old when they appear in *Current Contents*. An essential part of these publications is their subject and author indexes. There is also a authors' address directory and one for publishers too. The appropriateness of each publication to one's own information needs may be judged by scanning the list of journals covered, which is published twice a year. Retrospective searching, which would otherwise be a nightmare, is made somewhat easier by the appearance of a triennially produced cumulated index. The *Life Sciences* edition covers virology and provides contents lists for virtually all of the specialist virology journals. The *Agriculture, Biology and Environmental Science* edition supplements the data contained in the *Life Sciences* edition with its coverage of such journals as *Phytopathology* and *Veterinary Record*. For complete coverage of the virology literature both should be used.

Current Advances in Plant Sciences is a Pergamon publication of similar purpose. Published not quite so currently as the two above — monthly — it is a listing of titles from the current literature of plant sciences. Viruses of plants, on which about 30 titles appear in each issue, are covered under the subheading 'Plant pathology'.

Finally it must be said that the popularity of these current-awareness journals does not solely rest on their ability to report quickly on an article's publication; much of their success must be also attributable to their simple format and ease of use.

5.4 SUBJECT BIBLIOGRAPHIES

As conducting a literature search — an essential preliminary to any new research — has become easier thanks to the computer so there has been a growth in the number of bibliographies compiled and later published. (Bibliographies are here defined simply as lists that refer to writings related to a subject, period, author or other unifying concept. These lists provide sufficient details to enable an item's purchase or location.) Unfortunately the computer has led to bigger rather than better listings; too little thought is given to the implications of feeding in a particular keyword, with the result that many marginally relevant items are retrieved. In the 'old days' the compilation of a bibliography was considered to be a scholarly activity with each item being individually tracked down and assessed for its relevance; this rarely happens today, largely because of the size of the potential literature.

Two further drawbacks with subject bibliographies are: first, they may be highly personal in outlook; second, and more importantly, their 'one-off' nature means that they date exceedingly quickly — and what value, other than historical, is a list of documents published on, say, hepatitis in the 1960s in view of the fact that the literature has expanded enormously since then. Whilst there is no physical reason why a bibliography may not appear again in updated form, this seldom ever happens.

Because of the sheer number of bibliographies abroad the following discussion of individual bibliographies, conducted along subject lines, is necessarily selective.

Animal Viruses (*see* bibliography entries 333–344)

In the UK the staff of the Commonwealth Bureau of Animal Health in Weybridge have compiled a number of bibliographies on animal viruses based on review articles appearing in *Veterinary Bulletin*. These include *Rabies: epidemiology, pathology, immunology, diagnosis and control* (1978), containing 199 references from between 1976 and 1978, and *Bovine leukosis* (1978), containing 820 references from between 1972 and 1978.

In the USA the USDA Emergency Programs Foreign Animal Disease Data Bank have produced some very up-to-date bibliographies in topical areas. These include *Foot and mouth disease* (1980) with 4400 references; *Newcastle disease* (1978) with nearly 4000 references; *Hog cholera* (1981) with over 3000 references; *Ephemeral fever* (1981) with 300 references; *Visna maedi* (1980) with 600 references; and *Swine vesicular disease* (1980) with 660 references.

Bacteriophages (*see* bibliography entry 345)

The two volumes of *Bakteriophagie*, each in two parts, edited by Raettig cover phage research during the periods 1917–56 and 1957–65 and were published in 1958 and 1965 respectively. In all, 11 000 references are included and will presumably be of historical interest only.

Human Viruses (*see* bibliography entries 346–348)

The medical virologist is better served than his counterparts in the animal, bacterial and plant virus fields, although many bibliographies are old and have therefore been omitted.

The National Library of Medicine carries out searches of material listed in *Index Medicus* on request by individuals, which are then available as reprints, apparently at no charge. Reprints on virological topics include *Non A and non B hepatitis*, covering 1977–79 inclusive (1980) and containing 58 citations; and *Amantadine and influenza*, between 1975 and 1980 (1980) and comprising 114 citations. The searches are not restricted to medical subjects, an example being *Canine parvovirus*, 1978–80 (1980), comprising 44 citations. The NLM also produces, through its computer-based MEDLINE, a recurring list of citations of journal articles in specialized biomedical fields such as hepatitis.

On this subject, *A decade of viral hepatitis* compiled and edited by Zuckerman (1980) presents the great bulk of all hepatitis abstracts, amounting to about 1000 references for the decade between 1969 and 1979 selected from English-language journals.

In the field of tumour virology, the Biological Carcinogenesis Branch of the National Cancer Institute provide a regular computer listing of publications resulting from its collaborative research. It is known as the *Biological carcinogenesis research bibliography* and is updated in March and August of each year. It is available free to virus cancer researchers.

Methodology (*see* bibliography entries 349, 350)

Tissue culture and immunological techniques are indispensable tools in virology; thus it is appropriate to include some relevant bibliographies here.

Rapid and automated methods in microbiology and immunology: a bibliography 1976–1980 is compiled and edited by Palmer (1981). It contains approximately 3300 references divided into the various aspects of methodology such as immunofluorescence, purification, genetic techniques, cell culture, etc. It supersedes a similar publication covering the period 1967–75.

The enzyme-linked immunosorbent assay (ELISA) compiled by Voller *et al.* (1977) contains a review and a bibliography of microplate applications with 111 references and abstracts of papers on the ELISA, used for the detection of antigens and antibodies.

Plant Viruses (*see* bibliography entries 351–353)

As usual, plant virologists do not fare too well in terms of the number of bibliographies available, although the three that we are aware of are all relatively recent: *Bibliography of plant viruses and index to research* by Beale (1976), *Phycovirus bibliography*, compiled by Safferman and Morris (1977) and *The practical directory for phycovirus literature* by Safferman and Rohr (1979).

5.5 TRACING AND LOCATING VIROLOGY BOOKS

(*see* bibliography entries 354–361)

A motley collection of organizations combine (although not purposively) to list and record the publication of virology books. Amongst the organizations active in the field are libraries, government agencies, professional bodies, academic institutions and, perhaps the most numerous of all, commercial publishers.

Obtaining awareness of new books in the field is probably the most difficult task. This information reaches many people through the book review or books received columns of the journals they subscribe to or scan regularly. (Browsing through specialist bookshops or scanning the new-book shelves of specialist libraries are other popular methods.) Within the field of virology most journals take seriously their responsibility for reviewing books, or in the case of the less fortunate, listing them. Of the virology journals the most comprehensive treatment of books is provided by the *Society for General Microbiology Quarterly, Acta Virologica, ASM News* and *Zentralblatt für Bakteriologie, Mikrobiologie und Hygiene* — all of which review between 15 and 20 virology books (not necessarily the same ones) per year.

Whilst reading the book review columns of virology journals is rather a pleasant and painless way of keeping abreast of book publishing, there is no guarantee that all that is published will be reviewed, or indeed listed. Because of the sheer number of books coming out and the pressures on journal space, journals' coverage of the book literature is inevitably selective. In such a multidisciplinary field as virology, where the number of potentially relevant books is enormous, the Book Review Editor's problem is indeed a difficult one. Perhaps more serious than the problem of comprehensiveness is the one of currency. Thus most books are reviewed months, sometimes a year, after publication. Taking for instance the books reviewed by *The Society for General Microbiology Quarterly*, most of these have been published at least eight months before they are reviewed; some in fact are over a year old.

One way of minimizing, perhaps eliminating, the risk of missing a book review — and after all with a large number of journals as potential reviewers the risk is a real one — is to use a service such as *Index to Book Reviews in the Sciences*, which brings together the reviews published by a large number of journals. Published monthly by the Institute for Scientific Information, the *Index* lists by author book reviews published on average six months previously. The fact that the 1981 *Index* picked up a review of *The dictionary of virology* (1980) in *British Journal of Medicine* and one of *Virus and virus-like diseases of grapevines* (Bovey et al., 1980) in *Theoretical and Applied Genetics* shows how invaluable it is to the virologist who tends to read only the immediately related literature. In some months as many as 20 reviews of virology books or books on virology-related subjects are listed. As books of all scientific disciplines are listed together in the main author sequence, the index should be used to locate virology books. The actual review is not furnished, just sufficient details to enable its location. Other uses of the *Index* include: obtaining a number of opinions on one book and checking to see whether one's own book has been reviewed and by whom.

If a complete listing of new virology books is required then it is preferable to consult those reference tools which specialize in simply listing books. There are a great number of these available and only the most important are dealt with here. For British books there is undoubtedly no better source than the *British national bibliography* (*BNB*). The bibliography is classified, so most virology books can be found together, located at 576.4 (*see* section 5.8 for details of classification schemes); thus the relevant page serves as a mini subject bibliography. A subject index is provided which if searched under the terms *virus, virology* or the like will lead one to this section, or invaluably, if the book has been classified elsewhere, to that location also. A combined author/title index is also furnished. *BNB* lists about 15 virology books per year, amongst which may be counted reports, government publications and proceedings. It is a weekly service, listing books published between three and six months previously. Searching for new books can however be made irksome by the fact that it cumulates only at four-month intervals (Jan–April; Jan–Aug; Jan–Dec) and these cumulations can arrive several months late. Frequently then, at least six separate issues have to be consulted each time a search for new books is conducted. This problem can easily be overcome if *BNB* is searched on-line, where new data are automatically interfiled into one sequence; it is available via the British Library Automated Information Services (BLAISE). Other benefits of accessing the data via a computer are that it is available earlier than the hard-copy equivalent and more search pathways are provided, i.e. words in the title, publisher.

Because *BNB* has been going since 1950 it is also a source for older works. Again, as a large number of cumulations are involved (three-year cumulations cover the period 1950–70 and annual cumulations since then) it is much easier to search *BNB* on-line or on fiche. There is a 1950–75 cumulation available in this format, the value of which is diminished by the fact that it provides access by author or title only (so you need to know what you are looking for in the first place).

The best source for information on US books is the Library of Congress (LC). Its bibliographies are available in a number of forms. One of the most convenient, cheapest and most effective ways of obtaining LC data is through *MARCFICHE*.

MARCFICHE, one of the new breed of microfiche bibliographies, whilst not as generally available as other LC products, is well worth seeking out. Its value to the virologist lies in its: *currency* — it is a weekly service which frequently lists books within a month of their receipt by LC; furthermore the fact that the fiche can be sent airmail means that it can be obtained from the US very quickly indeed; *coverage* —LC buys virtually everything irrespective of language, subject, country of origin or form of publication (thus it is not just a list of US virology books nor does it confine itself to listing the main categories of books: theses, government publications, proceedings and research reports are included too); its *ease of use* — once you have obtained details of the book's accession number from the relevant index you simply consult the respective fiche (the index, which lists books published since 1964, is cumulated on a rolling basis so there is generally only one place to search); *classified index*, which lists material according to the Library of Congress Classification Scheme (*see* section 5.8) — the section of interest is QR–QY57, where over two hundred books are currently listed. A joint author/title index is also furnished. The major criticism that can be levelled against what is otherwise an excellent tool, is that the index provides only a very truncated form of the title (the first three words in many cases) and as a result it is not easy to discern from the index whether a book is really relevant or not. For amplification one has to go to the main numerical sequence.

The information available through *MARCFICHE* can also be obtained via an on-line search of the Library of Congress MARC tapes. In the UK these tapes may be accessed via BLAISE. Again the major advantage of such an approach is that books can be searched for in many more ways (i.e. any word in the book's title can be searched).

Abstracting and indexing services, whilst mainly in the business of drawing attention to journal articles, do also cover books, sometimes selectively: it is edited collections of papers or proceedings that tend to be singled out. *Virology Abstracts* is one that provides a comprehensive treatment of books, abstracting about 80 of them in 1981. Other abstracting/indexing journals that list books fairly comprehensively include *Excerpta Medica* and *Review of Plant Pathology*.

The advantage of listing in an abstracting service is that it is possible to make a good assessment of the book on the basis of the information contained in the abstract. The specialist book listing services mentioned above of course supply only sufficient information to enable a work to be traced; some information as to the quality and content of the work can be deduced from the publisher and title but this is minimal and incidental.

If it is currency rather than comprehensiveness that is required then it is publishers' catalogues or circulars that hold the key. Not only do they provide the earliest notification of a book's publication but they also go a considerable step further by providing up to six months' notice of the impending arrival of a new book. It is, then, to trade publications that we look for currency.

It is to the individual catalogues of specialist publishers that we look first. In the field of virology this means the catalogues of Academic Press, Blackwell Scientific Publications, the Cold Spring Harbor Laboratory, Springer and Elsevier. Some publishers' catalogues are fairly sophisticated affairs, with books arranged by a special classification (Oxford University Press) and fairly lengthy annotations provided (Elsevier). Academic Press is the biggest publisher of them all, with nearly 40 relevant titles in print. Its catalogue regularly reaches 300 pages in size and is broadly classified to enable browsing. Virology books are listed together with microbiology books. A title and subject index is also provided.

Circulars advising on the publication of new books are also forwarded to interested parties. Some of these are usefully presented in catalogue card form. Firms involved include Publishers Information Cards Services (PICS) and Blackwells, who will circulate to you, in the subject areas you specify, details of new and forthcoming books.

A complete list of publishers specializing in the biological, medical and veterinary sciences can be found in Cassell's and the Publishers Association's *Directory of publishing* (the 1983 edition is the most recent).

Because virology books may emanate from any one of a large number of publishers it is sometimes necessary to consult reference works that list together the entire product of the publishing trade. Two such lists — both in fiche and hard-copy form — are *British books in print* and the US equivalent *Books in print*. As books are listed alphabetically by author and title these works are outwardly of little value to those searching for virology books of which details are not known. However, both provide keyword entries, albeit rather haphazardly. Words are taken from the title and rely entirely upon the accuracy of the book's title in portraying its contents. If one looks under the terms *virus, virology, viral* for example, 85 references are listed in the US publication and 111 in its UK namesake (there is of course some overlap). The fiche edition of both should be preferred as it offers a monthly appraisal of virology books in print and new titles to come (three months' notice is commonly provided), whereas the hard copy provides such a service only on an annual basis.

The American *Books in print* has a sister publication that is probably of more use to the virologist: *Subject guide to books in print* is an annual publication which is essentially a rearranged form of the aforesaid bibliography. Books are listed under a large number of subject headings (based upon Library of Congress practice) which are themselves arranged alphabetically. Virology books are listed under the following headings: *veterinary virology; virology; virus research; viruses*; and under the names of specific viruses (a reference to the individual viruses represented is given under the heading *viruses*). About 300 virology books can be located under these various headings.

Two other popular bibliographies from the same stable as *Books in print* are deserving of comment: *Scientific and technical books and serials in print* and *Medical books and serials in print*. Both constitute relevant selections of documents from *Books in print* and *Ulrich's International periodical directory* and both are published annually in a handy one-volume reference form. The drawing together of serials and books on a topic has obvious attractions to the user. Their ease of use is considerably enhanced by the adoption of an alphabetical classified approach, which means that readers may alight quickly on virology books and serials. These two bibliographies, together with *Books in print*, may be accessed on-line via the DIALOG database.

A favourite and time-honoured way of learning about publications — though not necessarily new ones — on a particular topic is to check through the references that are so copiously appended to virtually every journal article or book. However this method might not prove too profitable as there is a marked tendency on the part of scientists not to cite books.

Finally there are a number of specialist bibliographies of virology publications in existence which, although heavily biased towards articles, do cover books also. These publications may be separately published or form part of a journal's issue. In this connection mention should be made of the bibliographies regularly issued by the Commonwealth Agricultural Bureaux and the United States Department of Agriculture (for more details of these publications *see* section 5.4 (animal viruses).

OCLC is a truly massive computerized database constituting some 9 000 000 bibliographic references to serials, monographs and other library materials. Unusually these data are available only via the on-line computer network. OCLC is centred on Ohio, although over two thousand libraries in the US and Canada are represented in the cooperative (libraries from European countries are also being encouraged to join). At present the value of OCLC is somewhat limited because it lacks a subject approach; however, an approach via words in a document's title can be made and, as mentioned on p. 120 in connection with serials, the provision of a subject search facility is planned in the near future.

.6 TRACING AND LOCATING CONFERENCE PROCEEDINGS (*see* bibliography entries 362–366)

To help overcome some of the problems of tracing conference proceedings there are a number of guides whose sole function is to list the proceedings that have been published. Most include not only separately published proceedings but also those that appear as part of periodicals. Howver it should be noted that: (a) many list proceedings and not individual papers, with the notable exceptions of *Conference Papers Index* and *Index to Scientific and Technical Proceedings*, and (b) they contain not the actual publications themselves or even abstracts of them but simply provide sufficient details to enable their tracing (the British Library's *Index of Conference Proceedings* offers the next best thing: a guaranteed loan or a photocopy of the required proceedings).

The *Index of Conference Proceedings Received by the BLL*, whilst being British-compiled, is international in coverage. Published monthly and cumulated annually, the *Index* lists the conference proceedings received by the British Library Lending Division, which is generally recognized to have the largest collection in the world (it currently receives about 15 000 a year). An 18-year cumulation (1964–1978) is available on microfiche, providing in one publication access to over 146 000 conference proceedings, making it the largest bibliography of its kind in the world.

In an attempt to overcome the problems of tracing proceedings the *Index* lists proceedings (alphabetically) by keyword extracted from the title, and sometimes the contents if the title is not sufficiently descriptive. As a consequence each proceedings may be listed several times under a number of different subject headings, thus increasing the chance of its location.

The adoption of an alphabetic subject keyword approach does mean that not all virological conferences are located together — they are scattered according to the alphabetical position of the words in their titles. A good number (14 in 1981) can be found under the terms *viral*, *virus* and *virology* but one also needs to look at the very least under the names of specific viral diseases and any other words of particular viral significance.

Most of the virology proceedings listed in the monthly issues are over a year old but this may be as much to do with late publication as late acquisition.

The *Index* is also available on-line via the British Library Automated Information Service (BLAISE) and is on the file known as *Conference Proceedings Index*. Nearly 150 000 conferences are listed on this file and at least 250 are virological in content. Apart from the speed of search the major advantage of accessing the *Index* via the computer is that it offers more search pathways to the enquirer. Thus a search may be conducted on any terms or combination of terms present in the conference's title (not just the keywords), author statement and publication description (date, location, etc.).

The *Index to Scientific and Technical Proceedings*, published monthly by the Institute for Scientific Information (ISI), is a relative newcomer to the field, having been started as recently as 1978 (it is therefore of limited retrospective use). The *Index* lists approximately 3000 proceedings from conferences held around the world, irrespective of language. Perhaps its major asset is that under each conference (arranged arbitrarily by an ISI accessing number) it lists all the authors and titles of the papers submitted — almost 100 000 papers are listed in this way. Six indexes provide a variety of approaches to locating a proceedings: by category of conference topics (broad subject headings, which include virology — on average 2 proceedings per month listed); by title words of papers, books, and conferences (about 60 can be located this way in any one month's issue); author/editor; sponsor; meeting location and corporate name. Reporting of conferences proceedings is relatively rapid, with the virology conferences being listed about 12 months late. There are exceptions however and not all the blame can be levelled at ISI. Thus a Conference on Genetic Variation of Viruses was held in New York between 28 and 30 November 1979. The proceedings were published by the New York Academy of Sciences in 1980 and eventually listed in the *Index* in November 1981 — two years in all. The value of such indexes in picking up conference papers published in unlikely spots is demonstrated by the listing of the proceedings of the Hepatitis Winter Workshop, published in *Medical Laboratory Sciences* (1981).

The monthly *Conference Papers Index*, published by Cambridge Scientific Abstracts, provides a classified listing of meetings, thus permitting browsing. There is a section called 'Biochemistry and biology' where most of the virology conference proceedings will be found. Under each meeting are listed all the papers that are featured in the published proceedings. Ordering details are also provided. Approximately 100 000 papers a year are indexed, of which around 4000 deal with viruses in one way or another. The subject keyword index should be used to locate virological papers presented at the conferences of other disciplines. *Conference Papers Index* is relatively current, with most papers having been presented at conferences held about six months previously. An annual index speeds retrospective searching, but it is probably easier, and certainly more effective, to use the on-line facility to conduct retrospective searches (the *Index* is available on DIALOG).

Two other specialist lists deserve attention, the first being Interdok's *Directory of Published Proceedings*, a chronologically arranged listing appearing monthly which is of considerable value to those searching retrospectively as it has been going since 1965. A keyword index provides access to virology conferences. Secondly, there is *Proceedings in Print*, which arranges proceedings alphabetically by conference name. Access to virology conferences is by the keyword index. This bi-monthly publication is mainly of value as a guide to purchase.

Finally, abstracting services should be mentioned, for after journal articles — which of course might be proceedings — proceedings normally obtain next best attention. *Virology Abstracts* offers the most comprehensive record of the proceedings of virology conferences. All proceedings listed in this journal are of course relevant so there is no question of sifting through mountains of index entries to proceedings of other disciplines. Unlike the previous services mentioned all conference papers are accompanied by a summary, which can amount to 80 words. Fifty conferences were covered by *Virology Abstracts* in 1981, most of these being reported nine months after publication.

5.7 TRACING AND LOCATING DISSERTATIONS (*see* bibliography entries 367–374)

Dissertation Abstracts International, published by University Microfilms, is probably the most widely known and available of the reference tools that provide details of recently accepted dissertations — in this case doctoral dissertations. It comes in three volumes, of which volume B: *The Sciences and Engineering* and volume C: *European Abstracts* are of interest. Very detailed abstracts (500 words) are furnished for each dissertation. Dissertations in volume B, published monthly, are mainly North American (South American, Belgian and Finnish dissertations are also featured) and are arranged under broad subject divisions to enable browsing. (Not all US universities allow University Microfilms to reproduce their theses, so *Dissertation Abstracts International* is not a source for all US dissertations.) Most, but not all, virology theses can be located under the biology subheading 'Microbiology'. Within this subheading they are arranged alphabetically by author, so virology dissertations are scattered amongst those of a microbiological theme. However, the subject index (there is also an author index) must be consulted, for about one-quarter of the virology theses can be located variously under 'Veterinary science'; 'Agriculture — plant pathology'; 'Health science'; 'Biophysics — medical'; 'Chemistry — agricultural biology'; and also under the general heading 'Biology'. The importance of dissertations as information sources (and the productivity of US universities) is underlined by the fact that *Dissertation Abstracts* currently lists around 200 virology dissertations per year. Most of these dissertations are abstracted about six to nine months after submission and most can be obtained in photocopy or microfiche form from the publishers, University Microfilms.

Dissertation Abstracts can be accessed on-line, together with *American Doctoral Dissertations* (a computer-generated index by keywords and author to US dissertations accepted between 1861 and 1972), *Comprehensive Dissertation Index* (an annual consolidation of US and Canadian dissertations listed in *Dissertation Abstracts International*) and *Masters Abstracts* (as the title suggests, a list of master's theses, primarily American) via DIALOG (in this form abstracts are not provided). The resultant database — COMPREHENSIVE DISSERTATION INDEX — provides a subject, title and author guide to virtually every dissertation accepted by an American university or college since 1861, when academic doctoral degrees were first awarded. Altogether the COMPREHENSIVE DISSERTATION INDEX contains details of three-quarters of a million dissertations, of which about 2600 are of virological interest.

For the UK, coverage of dissertations is the responsibility of two publications: *Index to Theses*, published by Aslib, and the British Library-issued *British Reports, Translations and Theses*. The Aslib index is a long-running one, having been published since 1950, and covers theses issued by colleges and polytechnics as well as universities. Again not all universities or colleges are represented, and the only way of learning about the theses produced by such institutions is to go to the individual universities' lists themselves. *Index to Theses* is published twice yearly, so cannot hope to be as current as its US counterpart. In fact theses are listed as long as 15 months after submission. They are arranged by broad subject heading, virology theses being listed in a subsection of the microbiology class. Yet again one has to turn to the subject index to retrieve all virology theses, for some may be located under other headings, most notably medicine. Around a dozen virology theses are listed per year, a far cry from the *Dissertation Abstracts* figure of 200. No summary of the content of each thesis is provided in the *Index to Theses*, merely sufficient data to enable it to be traced. However, abstracts of most of these theses may be obtained via its sister publication *Abstracts of Theses*, which is available only in microfiche.

British Reports, Translations and Theses is an invaluable tool as it is essentially a list of the British Library Lending Division's holdings of theses (this library receives all the theses listed in *Dissertation Abstracts*) and thus everything that is listed can be loaned or photocopied — and what is more, obtained within a few days. Published monthly, it is much more current than the Aslib list, with theses just 6–12 months late in being announced. As its name suggests, it is not just a list of theses, so one has to browse through reports and translations as well, which may of course be beneficial in that relevant items — and these publications are notoriously difficult to locate — may be picked up by accident. Material is arranged by broad subject, with virology being placed under 'Microbiology', which is itself a subheading of 'Biological and medical sciences' (06). Virology theses may also be located elsewhere in the arrangement, so the subject index must always be consulted. Approximately a dozen virology theses are listed per year. As the British Library has a collection of over 20 000 theses it is well worth searching retrospectively (via the annual indexes) through *British Reports, Translations and Theses*. It is possible to go back as far as 1971.

Some abstracting and indexing journals (viz. *Virology Abstracts*) also cover theses, but usually selectively.

5.8 CLASSIFICATION SCHEMES USED TO ARRANGE VIROLOGY DOCUMENTS IN LIBRARIES AND BIBLIOGRAPHIES (*see* bibliography entries 375–380)

Finding out where a book, proceedings, report or periodical is located in a library, catalogue or bibliography can prove to be a nightmare, particularly in the case of the library. The experience can be enough to dissuade many people from ever bothering to understand why a document is in a particular place. It is largely for these people — perhaps the majority of virologists — that this section has been written. Thus a very pragmatic attitude to the whole business has been adopted: first, the precise locational marks at which virology material may be found in the major classifications used by libraries and bibliographies are pointed out, and secondly, the reasons for the particular location are outlined. It is an irony of classification that by grouping one inevitably also scatters (and it is often virology documents that are scattered).

Classification schemes, of which there are unfortunately too many, exist primarily to arrange documents — whether on library shelves or on the pages of bibliographies — in a manner that has been held to be the most helpful to the user: by subject. The principle underlying classification is, in theory anyway, simple: documents that are judged to be on the same or similar subjects are gathered together. The subject path to information is the most trodden because it offers an approach to those (and that is most of us, most of the time) who do not know precisely what they are looking for in the first place, but have only a rough idea of what is available. It is then, by providing a framework in which the searcher can browse and make comparative assessments of related material, that classification is so helpful.

Unfortunately a number of factors complicate the whole process, not least being that there exists little agreement as to precisely what particular subject arrangement (and there is a wide choice) would be most useful. In a field such as virology for instance we have a number of user groups — biologists, veterinary scientists and doctors, to name the main parties — who all view the field from a fundamentally different stance; and all these viewpoints translate into quite different arrangements of documents on library shelves and in the pages of bibliographies and related tools. *Table 5.3* demonstrates this point by showing how the same publication can be treated quite differently by the major classification schemes. Book 1, for instance, on rabies is classified under 'Microbial diseases' by Dewey, under 'Medical sciences' by the Library of Congress and under 'Virology' by Barnard.

Secondly, subjects as diffuse as virology are not easily compartmentalized and assigned places in what after all is a linear sequence. Such a sequence cannot of course accommodate the multifarious and multidisciplinary relationships virology, for one, possesses. Thus the central problem that bedevils all virology classifications is whether documents on viruses should be lodged together with books covering the other diseases that infect a particular host — man, plant or animals (the approach favoured by the Dewey Decimal Classification scheme) — or whether all books on viruses, regardless of their host, should be placed together (the approach favoured by the Library of Congress). It is easier to tackle this problem in catalogues and bibliographies, where a reference to a document can be put in a multitude of places (i.e. wherever that document has a relationship — although because of publishing economics this is seldom done), than in libraries, where one would have to buy multiple copies of the document and shelve them in all the appropriate places (a very costly business).

Thirdly, subjects, particularly those in science, are continually expanding their horizons — at one and the same time strengthening and weakening subject relationships — and as a result forming new groupings of knowledge. Virology itself has only recently emerged from such a process to become a subject in its own right and naturally, with one so recently born, contains many of the vestiges of its parental subjects — medicine, veterinary science and biology. Indeed, much of

Table 5.3 Different approaches to classifying books: Dewey, LC and Barnard

Book	Scheme		
	Dewey	LC	Barnard
1. Rabies (Bisseru)	576.2 Biology—Microbes	RC 148 Medicine—Infectious Diseases	K.K. Virology—Rabies
2. Plant virology (Matthews)	576.64 Biology—Microbes—Viruses	SB 736 Agriculture—Viruses of Plants	K.Y. Virology—Plant Viruses
3. Diagnostic procedures in virology (Lennette)	576.62 Biology—Microbes—Rickettsiae	QR 46 Medical Microbiology	K.P.Q. Virology

the subject still resides in the literature and organizations of these disciplines. Few classification schemes can keep abreast of such rapid and widespread change and as a consequence reflect positions as they were years, perhaps decades, ago. Thus 'Dewey' groups viruses with rickettsiae under 'ultramicrobes'; this is very antiquated as the bacterial nature of rickettsiae is well established. However, it is not only libraries and bibliographies that find themselves in difficulties, for the field itself is, unusually for science, plagued with difficulties over taxonomy and nomenclature. Thus after the third report of the International Committee on Taxonomy of Viruses (1979), the President concluded that the taxonomy of viruses is really at the same general position as the taxonomy of plants and animals was in the period from 1700 to 1859! The situation is so bad that no official names for individual viruses have yet been approved.

Unfortunately then, because of the factors outlined above the student, teacher and researcher seeking information on a particular subject are highly likely to encounter variant, outdated and sometimes confusing classification practices. The wide range of classification schemes likely to be met with is illustrated in *Table 5.4*.

Table 5.4   Classification schemes in the UK

Own scheme	33%	(122)
NLM	14.9%	(55)
Barnard	14%	(52)
Dewey	13.5%	(50)
UDC	13.8%	(51)
LC	5.4%	(20)
Bliss	3%	(11)
Others — Boston, MRC, Garside, Bodleian — less than 1% each		

Based on an examination of 370 specialist libraries in the UK.

Dewey Decimal Classification

'Dewey', as it is affectionately called, is probably the most widely encountered classification scheme. In a vain attempt to keep abreast of the ever-changing structure of knowledge it has run to 19 editions. Unfortunately its primary structure was laid down by Melvil Dewey over a century ago and whilst changes have been made (largely to accommodate the growth of science) it still retains much of its original structure.

Since the Decimal Classification (DC) scheme encompasses the whole of knowledge, virology merits only a very tiny space in the scheme — at 576.64 to be precise (*Table 5.5*). The placing of viruses here is both antiquated and inconsistent, since, as mentioned above, the bacterial nature of rickettsiae is well established. Admittedly, their size is about the same as that of large viruses, but

Table 5.5 The whereabouts of virology in the Dewey Decimal Classification system

Notation	Subject
500	PURE SCIENCE
570	Life sciences
574	Biological sciences
.2	Pathology
575	Organic evolution and genetics
576	Microbes
.6	Ultramicrobes
.61	Types
.62	Rickettsiae
.64	Viruses
.648	Types
.648.2	Bacterial viruses
.648.3	Plant viruses
.648.4	Animal viruses
577	General nature of life
580–9	Botanical science
590–9	Zoological science
600	TECHNOLOGY. APPLIED SCIENCE
610	Medical sciences
616	Diseases
630–9	Agriculture

then so too are chlamidae and mycoplasmas, which are correctly classed as bacteria. Virology at 576.64 is further subdivided into bacterial viruses (576.6482), plant viruses (576.6483) and animal viruses (576.6484). Within these categories viruses are listed alphabetically. This is not an entirely satisfying solution as numerous synonyms exist for individual viruses (e.g. infectious bursal disease/Gumboro disease virus) and doubts obviously arise as to the name under which the documents are stored. Additionally, it would have been preferable first to group animal viruses into vertebrate and invertebrate viruses and to have provided under plant viruses a separate group for fungal viruses.

Although the idea, adopted by the Dewey scheme, of ordering viral groups according to the nature of the host that is infected is contrary to the policy of the International Committee on Taxonomy of Viruses (which prefers a classification based upon the characteristics of the virus itself), it does reflect current practice, for most virologists work with viruses affecting only one kind of host.

Whilst it would seem from the above that virology publications are largely collated together and may be found in one place on library shelves using Dewey, this is far from the truth because Dewey places documents according to the discipline in which they are studied. *Table 5.6* illustrates this point nicely. Of the 12 books published on viruses between 1975 and 1979 only 4 are placed at the

Table 5.6 Scatter of virology publications throughout the Dewey Decimal Classification

1. 368.4'1 — Man, Liver, Viral hepatitis. Health insurance benefits. Provision. Great Britain. Inquiry reports.
    Industrial Injuries Advisory Council
    Viral hepatitis: report/by the Industrial Injuries Advisory Council in accordance with Section 141 of the Social Security Act 1975 on the question whether viral hepatitis should be prescribed under the Act.

2. 576'.64 — Picornaviruses. Molecular biology. Conference proceedings.
    NATO International Advanced Study Institute on the Molecular Biology of Picornaviruses, Maratea, 1978. The molecular biology of picornaviruses/ edited by R. Perez-Bercoff.

3. 576'.64 — Viruses
    Primrose, S. B. Introduction to modern virology.

4. 576'.64 — Viruses. Reviews of research.
    Advances in virus research.

5. 576'.6483 — Plants. Pathogens: Viruses.
    Matthews, Richard Ellis Ford. Plant virology.

6. 582'.16'0234 — Trees. Virus diseases.
    Cooper, Joseph Ian. Virus diseases of trees and shrubs.

7. 596'02 — Vertebrates. Tumours. Role of viruses — Conference proceedings.
    Munich Symposium on Microbiology (5th : 1980)
    Leukaemias, lymphomas and papillomas: comparative aspects / (proceedings of the 5th Munich Symposium on Microbiology, held on 3–4 June 1980 and organized by the WHO Collaborating Centre for Collection and Evaluation of Data on Comparative Virology); edited by Peter A. Bachmann.

8. 599'08'765 — Mammals. Cells. Plasma membranes. Effects of oncogenic viruses.
    Virus transformed cell membranes/edited by Claude Nicolau.

9. 616.01'94 — Man, Pathogens: Viruses. Conference proceedings.
    Persistent viruses/edited by Jack G. Stevens, George J. Todaro, C. Fred Fox.

10. 616.01'94'028 — Medicine, Virology, Diagnosis, Laboratory techniques.
    Diagnostic methods in clinical virology.

11. 616.9'94'0194 — Man. Cancer. Pathogens: Viruses. Control measures. Conference proceedings.
    NATO International Advanced Study Institute on Antiviral Mechanisms for the Control of Neoplasia, Corfu, 1978. Antiviral mechanisms in the control of neoplasia/edited by P. Chandra.

12. 633.63'98 — Sugar beet plants. Beet mild yellowing virus and beet yellows virus.
    Agricultural Development and Advisory Service. Virus yellows of sugar beet.

Source: *British National Bibliography*

regular virology number 576.64 or its subdivisions. The others are variously listed in the social service class (book 1), botanical class (bk 6), zoology class (bk 7), medicine class (bks 8, 9, 10, 11) and agriculture class (bk 12). Furthermore, this is not an exhaustive list of all possible locations.

Finally, one further small criticism of Dewey: interferons are included under virology (576.64) to reflect the viral relationship of these proteins but would be better classified with immunology as they represent host defence responses to virus infections.

Universal Decimal Classification

A variant of the Dewey scheme, the Universal Decimal Classification (UDC) differs in that it offers far more specific identification of subjects and a more flexible approach to positioning documents. It suffers from the fact that to achieve this it has had to resort to very long numbers (10 — far too many to remember — is not uncommon) and rather confusing punctuation. UDC is primarily used in special libraries and collections and in international bibliographies (it is supposedly an international scheme, free from national bias).

A long-awaited revision of the Biological Sciences section was prepared by the British Standards Institution (the publishers of UDC) in 1979, superseding material last compiled in 1943. The two editions differ remarkably in respect to their treatment of virology. The 1943 edition (which may still be in use in some libraries and bibliographies — the costs of reclassifying being prohibitive) places virology somewhat inconspicuously under 567.858 whereas the new edition recognizes the new-found importance of the subject by placing it on an equal footing with bacteriology, and classifies it at 578 (*Table 5.7*).

In fact the 1947 edition was published only a few years after a virus had been seen for the first time, and not very clearly at that, so it is quite understandable that it should base its classification on the tissues infected or the effects caused. After some 40 years there is no need to stress the obsolescence of the terminology used. Some libraries still using the old edition arrange all virology books at 576.858 alphabetically by either author or virus. Such an arrangement of course does not aid swift retrieval.

The new edition is virtually a new scheme. Its prime aim is to bring together the whole of virology under one heading and thus overcome the scatter inherent in Dewey's approach. The result naturally is a rather indigestible notation (e.g. Rabies 578.824.11). The arrangement and nomenclature adopted follow the proposals and recommendations of the International Committee on Taxonomy of Viruses: viruses are grouped according to host and listed using latinized binomial names grouped mainly according to genera (ending ... virus), subfamilies (ending ... virinae) and families (ending ... viridae). Every major virus obtains its own place in the scheme. For example, rabies virus is classed as follows:

578.82	Viruses primarily of vertebrates
578.824	Rhabdoviridae
578.824.11	Rabies virus group

Note too how the scheme, like Dewey, is essentially hierarchical in nature with

Table 5.7 The whereabouts of virology in the Universal Decimal Classification

Notation	Subject
1979 edition	
57	Biological sciences in general—virology and microbiology
.083	Microbiological, virological and immunological techniques
.083.2	Virological methods and techniques
.083.24	Isolation of viruses
578	*Virology*
.1	Viral biochemistry
.2	Molecular virology
.3	Morphology of viruses
.4	Viral ecology
.5	Viral genetics
.56	Viral genetic exchange and recombination
.7	Medical virology
.8	Classification and systematics of viruses
579	Microbiology
.61	Medical microbiology
.62	Veterinary microbiology
.64	Agricultural microbiology
616	Diagnostic medical microbiology and pathogenicity of microorganisms
619	Veterinary science
631	Agriculture
1943 edition	
576.8	Microbiology, bacteriology and parasitology
576.858	Virus species. Filterable and ultramicroscopic bacteria
.1	Dermotropic virus spp.
.2	Neurotropic virus spp.
.3	Myotropic virus spp.
.4	Epitheliotropic virus spp.
.5	Adenotropic virus spp.
.6	Tumour-forming viruses
.7	Catharral virus sp.
.8	Mosaic viruses of plants
.9	Bacteriophages

the notation reflecting and supporting this characteristic. Inevitably some scatter of virological material still occurs but where this does happen it is well signposted.

The overall position of virology on the library shelves, between material on the 'Basis of life' (577) and 'Microbiology' (579), is a good one as it emphasizes the ambivalent position viruses hold, in coming somewhere between the living and the non-living.

The only (small) flaw in the scheme would seem to be the separation of information about virological techniques (57.083.2) from the main body of the literature at 578.

Library of Congress Scheme

The Library of Congress (LC) scheme, found mainly in university libraries, was designed in 1897 and is now into its 6th edition (1973). It is similar to DC and UDC in that it encompasses the whole of knowledge but differs in that it adopts a mixed notation of capital letters (designating all the major disciplines) and numbers (indicating the subclasses of these disciplines) (*Table 5.8*).

The scheme attempts to bring together the veterinary, medical and biological aspects of virology under one subject heading and this it does with mixed success. Within this arrangement material is collated by host and disease rather than by virus. Consequently there is a great deal of ambiguity as to where individual topics might be located. For example, smallpox could theoretically be found under:

QR	160	Microorganisms of the human body
	189.5.56	Vaccines
	201.56	Pathogenic organisms
	412	Pox viruses
RC	110	Medical infectious diseases

as could many other viruses. Signposting is however used extensively and helps to offset the scatter (this of course is a facility not offered to those browsing the shelves).

The main core of virology is found under QR.355. Under 'Systematic divisions' (396) will be found ten of the main groups of viruses with their own specific entry, e.g. adenoviruses (QR 396), herpesviruses (QR 400). The remainder are amalgamated in alphabetical order under QR 416. Alphabetical listing is a feature of the scheme. Vaccines and pathogenic microorganisms are similarly treated.

The bacterial viruses or bacteriophages are mysteriously excluded from the bulk of virology and placed at QR 342, being classified as a microorganism of microorganisms!

Generally speaking microbiology fits uncomfortably between experimental pharmacology (QP 901) and comprehensive medical texts (R). More relevant bedfellows, genetics (QH 426–470), cytology (QH 573–671) and biochemistry (QP 501–801), are diffusely and distantly placed. A survey of how a dozen virology books were classified showed that it was the LC scheme which scattered the most widely; 75% were given numbers other than the virology ones.

Table 5.8 The whereabouts of virology in the Library of Congress Classification scheme

Notation		Subject
Q		SCIENCE
QH		Natural history, biology (general)
QP		Physiology
QP 901		Experimental pharmacology
QR		Microbiology
QR	46	Medical microbiology
QR	49	Veterinary microbiology
	65	Techniques
	160	Microorganisms in the human body
	175	Virulence
	180	Immunology
	189	Vaccines — by disease A–Z
	201	Pathogenic organisms A–Z
	301	Microorganisms of animals
	342	Bacteriophages
	351	Microorganisms of plants
	355	*Virology*
	355–372	Reference, periodicals, etc.
	385	Techniques
	393–394	Nomenclature, taxonomy
	396	Systematic divisions
	450	Virus structure
	465	Virus function
	480	Virus interactions
R		MEDICINE
RB		Pathology
RC		Internal
	110–253	Infectious diseases
	666–923	Systematic diseases
S		AGRICULTURE
SB		Plant culture and horticulture
	733	Microbial diseases of plants
	736	Virus diseases of plants
	942	Diseases of insects
SF		Veterinary science
	781	Virus diseases of animals

Barnard Classification

While the three schemes mentioned previously attempt to embrace all of knowledge, the Barnard Classification is a specialist scheme addressing itself to only part of knowledge. The Barnard Classification was designed with medical and veterinary libraries in mind. It was first published in 1936 and last revised and expanded in 1955. The principle underlying the scheme is that of specific entry; that is, one place for each topic under which are grouped all its aspects. Its notation is primarily alphabetical with just the geographical subdivisions of classes being indicated by arabic numerals. The 1955 revision recognized the importance of virology by designating it a class on its own — class K (*Table 5.9*). Previously it had shared a class with bacteriology.

Table 5.9 The whereabouts of virology in the Barnard Classification

H	Immunology and infectious diseases
HA	Infection
HB	Immunity
HG	Serology
HI	Immunization
HW	Infectious diseases in general
J	Bacteriology
K	Virology and viral diseases
KA–KF	Rickettsial typhus group
KG	Psittacosis
KH	Insect viruses
KI	Pox group
KJ	Herpes group
KK	Rabies
KL	Influenza group, mumps, etc.
KM	Rous sarcoma
KO–KV	Specific viruses largely based on clinical criterion
KW	African animal viruses
KX	Other viruses of man and animal
KY	Plant viruses
K2	Bacteriophages
O	Transmission of communicable diseases
P	Pathology (tumour viruses)
Q	Diagnosis and clinical medicine
RK–RV	Therapeutics, etc.
X	Veterinary science

It is to be expected that a highly specialist scheme such as this will inevitably result in a relatively greater physical scattering of literature than a general one by virtue of the greater volume of books contained by the library covering interdisciplinary subjects. With this system it is possible to locate individual viruses; for example, laryngotracheitis of fowls is KXL. It is possible however that more documents might also be found under 'Diseases of poultry', class C.XUD. A major fault of the scheme, from the point of view of virology, is the inclusion of tumour viruses, on which there are numerous books, under 'Pathology' and not under 'Virology'.

There is a systematic schedule of general subdivisions for use within classes, divisions or sections; for example, AX implies recent advances in ... . Auxillary schedules may apply for use with any disease; thus J refers to a viral causative agent of a particular disease and there are other notations to distinguish types of tumour and control of viral diseases.

Because of its inflexibility, largely resulting from an outdated outlook, this scheme is obviously in need of complete revision; also, rickettsiae and psittacosis need to be removed from the viruses. The nomenclature is antiquated and the classification is haphazard, with no taxonomic significance, and, at times, erroneous.

National Library of Medicine Scheme

The National Library of Medicine system is a development of the Library of Congress scheme with the difference that it allows for much more specific entry, thus enabling relatively small or narrow sub-fields of virology to have their own places in the scheme. The latest edition (4th) of the scheme was published in 1978, so it is relatively up to date. The scheme will be met mostly in searching MEDLINE, the giant computer abstracting service.

The scheme naturally has a medical bias (*Table 5.10*) and is divided into 'Preclinical sciences' (QS–QZ) and 'Medicine and related subjects' (W–WD). The greater part of virology will be found between QW 160 and QW 170, where a virus or groups of viruses can be specifically located, e.g. 165.5 A 3 Adenoviruses. Immunological aspects of virology, such as vaccines, will be found between QW 500 and QW 949. The other location for viral literature is 'Infectious diseases' (WC 207–950) and a minimal amount may also be located under 'Pathology' (QZ 65).

5.9 EPILOGUE

Much information of interest to virologists is held on computer — some of it exclusively so. The on-line services offered by such database hosts as DIALOG and ORBIT, which between them offer access to around 200 published reference and bibliographic tools, go a long way to making information more available and take out much of the slog associated with retrieving information. It is perfectly feasible to use these databases with, say, a day's training, but it is more normal to avail oneself of the search services offered by a library or information unit. Surprisingly, the costs of a computer search compare very favourably with those of a manual search, for one must set against the seemingly high hourly cost (£30–£70) what a computer can actually do in an hour.

On-line services such as DIALOG are now offering a document delivery service, by which relevant items may be forwarded automatically by post to the user. This offers to the users the tantalizing prospect of never stepping in a library! Subject

Table 5.10 The whereabouts of virology in the National Library of Medicine's classification scheme

QV		PHARMACOLOGY
QW		MICROBIOLOGY AND IMMUNOLOGY
	1	Societies
	4	General works
	25	Laboratory manuals
	70	Veterinary microbiology
	160	Viruses (general), virology
	160.3	English language 1978–
	160.4	Other languages 1978–
	161	Bacteriophages
	161.5	Specific phages A–Z
	162	Insect viruses
	164	Vertebrate viruses
	165	DNA viruses
	165.5	Specific DNA viruses A–Z
	166	Oncogenic viruses
	167	Oncolytic viruses
	168	RNA viruses
	168.5	Specific RNA viruses A–Z
	169	Vertebrate viruses, unclassified
	170	Hepatitis viruses
	180	Pathogenic fungi, mycology
	500	IMMUNOLOGY
	700	Infection, mechanism of infection and resistance
	730	Virulence
	800	Biological products producing immunity
	805	Vaccines, etc.
	806	Vaccination
	815	Immune serums
	949	Vaccine therapy
WC		INFECTIOUS DISEASES
	207	Viral pneumonia
	500	Virus diseases (general or not indexed elsewhere)
	505	Viral respiratory tract infection (as above)
	585	Specific diseases such as influenza, measles, smallpox
	950	Zoonoses
QZ		PATHOLOGY
	65	Bacteria, fungi, viruses, rickettsiae (general)
QX		PARASITOLOGY

searching can be done remotely (even from one's own home) and items forwarded to any address one might wish.

References and Further Reading

Cho, Yong-Ja (1976). Usage of periodical literature in veterinary science. In *4th Canadian Conference on Information Science, University of Western Ontario, 1976.*

Sengupta, I. N. (1974). Choosing microbiology periodicals: study of the growth of literature in the field. *Annals of Library Science and Documentation* **21**(3) : 95–111.

6 Culture collections

6.1	Importance of culture collections	161
6.2	Culture collections and patents	163
6.3	Culture collection catalogues	164

IMPORTANCE OF CULTURE COLLECTIONS

Culture collections are to the microbiologist what animal and plant specimens are to the zoologist and botanist. Onions (1979) wrote of culture collections, 'Without them the science of microbiology can neither be taught effectively nor developed efficiently for industrial or research purposes. They are essential for the progress of microbial taxonomy and to support accurate identification of microbes upon which, ultimately, all microbiological science depends.'

In the UK there are some 32 collections devoted to bacteria, fungi and yeasts but none to animal, insect or plant viruses. Five bacteriophage collections exist although with names that somewhat disguise their function: the National Collection of Dairy Organisms, the National Collection of Plant Pathogenic Bacteria, the National Collection of Marine Bacteria, the National Collection of Industrial Bacteria and the Rothamsted Rhizobium Collection. The interest in phages, here, is mainly for 'typing' the bacteria rather than for the viruses themselves.

Writing in the *United Kingdom Federation for Culture Collections Newsletter*, Magrath (1976) suggested that the lack of a National Viral Collection is due to the relative expense of preparing and storing stocks of virus. He classifies virus strains into:

1. Strains used for vaccine production
2. Strains with special properties making them of interest in research projects
3. Strains without a detailed pedigree
4. Prototype strains used for classifying unknown viruses

The availability of vaccine strains is often restricted but they may be obtained from manufacturers or national control authorities, who will have details of their pedigree. Strains having special properties are best obtained from the individual publishing the description of the particular characteristics of that strain. The third group, the 'mongrel' strains, are unlikely to be in any great demand. It is within the fourth group, the prototype strains, that the need for a culture collection exists.

Identifying unknown strains of viruses with prototype viruses also requires strain-specific sera and only a laboratory committed to the preservation of viruses and sera could be expected to maintain them indefinitely and keep details of passage history. The Central Public Health Laboratory at Colindale, the Centre for Applied Microbiology and Research at Porton Down and the Animal Virus Research Institute at Pirbright partly fulfil the function of culture collections but their selection of strains is limited.

In the USA, where there are a number of virus collections, the most important is the American Type Culture Collection (ATCC) at Rockville. The ATCC is an independent non-profit-making organization devoted to the preservation of reference cultures of microbes, viruses and cell lines and to their distribution to the scientific community. According to its 1979 Annual Report (1980), its current virus holdings then comprised 1050 strains of vertebrate virus and 200 antisera to vertebrate viruses (including viral reagents catalogued and distributed by the National Institute for Allergy and Infectious Diseases (NIAID), Bethesda), 215 plant viruses and 67 plant virus antisera.

A survey of culture collections on a worldwide basis was carried out in 1966 by the IAMS Section on Culture Collections, now the World Federation for Culture Collections, on the instructions of Unesco. As a direct result, a *World directory of collections of cultures of micro-organisms*, compiled and edited by Martin and Skerman (1972) was published. As well as Unesco, the WHO and CSIRO were major financial contributors. The primary function of this directory is to provide a means whereby a particular species of microorganism or a cell line can be located. It has a geographical index, an index of collections listing important details such as whether catalogues exist, full address, main interests (e.g. medical microbiology, veterinary microbiology) and an index of microorganisms that also lists the common names of animal viruses. The number of virus cultures listed in the *Directory* is as follows:

Animal viruses	58
Bacterial viruses	50
Insect viruses	4
Plant viruses	12
Tissue culture	6

In the present volume, the names and addresses of major culture collections are given as directory entries 530–551.

References

Magrath, D. I. (1976). The need for a national collection of virus strains. *United Kingdom Federation for Culture Collections. Newsletter* No. **2** (May 1976).

Martin, S. M. and **Skerman, V. B. D.** (Eds) (1972). *World directory of collections of cultures of micro-organisms*. New York: Wiley-Interscience.

Onions, A. H. S. (1979). National culture collections. *Society for General Microbiology Quarterly* **6** : 134–6.

6.2 CULTURE COLLECTIONS AND PATENTS

While microbiological processes and products have a long history of patents, mainly in the area of applied science such as vaccines, antiviral agents, antibiotics and techniques, it was only in 1980 that a Supreme Court decision in the USA declared a living organism to be a patentable subject matter. Similar developments, inevitably, will follow in Europe and the rest of the world, although the sheer size of the American market makes this a significant step in its own right.

A particular requirement of patent law, as it stands in 1981 in the UK and the USA, is that the description of patent specification should be sufficient to allow a skilled expert to repeat the experimental procedure. Arising from this, the microorganism, whether being described either as part of a process or as itself, must be deposited in a culture collection prior to the date of filing the application. Other bodies insisting on deposition, up to 1979, include the Japanese, Canadian, Dutch, West German, Hungarian and French Patent Offices and the European Patent Office (EPO).

The US Patent Office officially recognizes the ATCC as a public repository for microbial strains involved in patent applications. For its part, the ATCC regularly accessions, maintains and distributes, where legally permissible, cultures of animal and plant viruses, cell lines, recombinant DNA, plasmids and non-viral microorganisms for patent purposes. This amounted to 2000 patent cultures in 1980.

In 1971 the US Patent Office published guidelines governing the deposition of microbial cultures for patent purposes. These and guidelines issued by the EPO and the World Intellectual Property Organization (WIPO) as well as others serve as the basis for ATCC letters of agreement with depositors of patent cultures.

Deposition in a US culture collection is recommended if an applicant, no matter what his or her nationality, is likely to want protection in the USA at some time because US Court decisions have held that deposit in a non-US culture collection cannot be proven. The Budapest Treaty on the International Recognition of the Deposit of Micro-organisms for the Purpose of Patent Procedure (1977) will help to reduce this problem because a deposit in a single culture collection acceptable to all ratifying members will be valid for patent applications in those countries. Such ratification is slow in coming, and until it is unanimous, the US collections must be considered to be the major depositories for the microorganisms named in UK patents.

An annual fee is levied for deposition, and this affords protection for 17 years, although in fact international treaties stipulate that the culture must be maintained for 30 years even if the patent is abandoned. For animal viruses or cell lines at least 20 ampoules of the culture should be deposited. Viability tests must be performed on the culture on deposition as required by WIPO and EPO treaties.

Granting of a patent requires full disclosures of the invention, thus giving scientists access to the most advanced developments in the field. In return the government grants the inventor the right to exclude others from making, using or selling that invention, usually for 17 years.

6.3 CULTURE COLLECTION CATALOGUES

Catalogues can form a useful quick reference source on microorganisms. The American Type Culture Collection (ATCC) catalogue, which issues a new edition every two years, gives considerable details of depositor, patents and references to the strain in the journal literature. Bannister and Oppenheim (1979) found that this was not true of a number of catalogues of British culture collections such as the National Collection of Dairy Organisms and the National Collection of Yeast Cultures. These were often unreliable and scanty as regards information.

Other collection catalogues of interest to virologists are as follows:

(a) *Catalogue of cultures in the National Collection of Plant Pathogenic Bacteria*. This is published and compiled in the UK by the Ministry of Agriculture, Fisheries and Food (1977) and lists bacteria and phages maintained by the UK's national collection. (Phages useful for the determination of bacterial plant pathogens are also maintained.) The catalogue contains details of how to order the specimens and an appendix listing the English common-name equivalents of plant hosts.

(b) *National Institute for Allergy and Infectious Diseases — catalogue of research reagents 1978–1979.* This is published by the US Department of Health, Education and Welfare and includes microbial reagent data, many of them viruses: picornaviruses, adenoviruses, hepatitis A and B, arboviruses and interferons.

References and Further Reading

American Type Culture Collection (1980). Annual Report, 1979. ATCC, Rockville, Maryland.

Bannister, D. (1979). British microbiological patents. *Society for General Microbiology Quarterly* **6** : 105–11.

Bannister, D. and **Oppenheim, C.** (1979). Information about microorganisms contained in patent specifications. *Journal of Chemical Information and Computer Science.* **19** (3) : 123–5.

Cox, R. B. (1980). EEC patent law rule change. *Society for General Microbiology Quarterly* **7** : 129.

Jones, D. (1979). National culture collections. *Society for General Microbiology Quarterly* **6** : 136–8.

Martin, S. M. and **Skerman, V. B. D.** (Eds) (1972). *World directory of collections of cultures of microorganisms.* New York: Wiley–Interscience.

7 Legislation and laboratory safety

7.1 UK legislation and codes of practice	165
7.2 US legislation and safety rules	167

7.1 UK LEGISLATION AND CODES OF PRACTICE

In the UK, a wide variety of legislation affects the virologist, primarily because he or she is a laboratory worker and thus subjected to the risks and responsibilities that such work imposes, and secondly because of the special nature of the organisms he or she works with.

Like other biologists, the virus research worker faces the highly contentious Cruelty to Animals Act 1876 whereby the Home Office, in the form of licences and inspections, exercises administrative control over the use of experimental animals. A survey of the law relating to laboratory animals is given by Cooper (1980). More detailed information can be found in *Guidance notes on the law relating to experiments on animals in Great Britain*, which is issued by the Research Defence Society and compiled in collaboration with the Home Office (1974 and 1976).

Much attention has been paid lately to safety in laboratories, and, in particular, to the prevention of laboratory-acquired infection. A number of significant developments have taken place in this area: for example, the passing of the Health and Safety at Work Act 1974 and the *Report of the Working Party into the Laboratory Use of Dangerous Pathogens* (the Godber Report, 1975). Following Godber's recommendations, *A code of practice for the prevention of infection in clinical laboratories and post-mortem rooms* (Howie, 1978) laid down increased health and safety standards of equipment and procedures for dealing with infectious microorganisms in clinical, research and teaching laboratories. Though the Howie Report, as it has become known, is not strictly legally binding, the fact that the Report is applied by the Health and Safety Executive of the DHSS as criteria for judging compliance with the Health and Safety at Work Act 1974 gives it particular importance. For the purpose of the Code of Practice, as recommended by Howie, viruses and other microorganisms are classified into four groups depending upon the degree of hazard they present. It then describes the minimal safety conditions for handling them.

Category A organisms include the most pathogenic known to man. Endorsement is required by the Dangerous Pathogen Advisory Group of the DHSS in order to work with them. All these organisms are viruses.

> Simian herpes (BO) virus
> Lassa fever virus
> Marburg virus
> Rabies virus
> Smallpox virus
> Crimea (Congo) haemorrhagic fever virus

Conditions are also laid down to cover the accidental isolation of these viruses in clinical laboratories.

Category B1 organisms include viruses that offer special hazards to laboratory workers and for which special accommodation and conditions for their containment must be provided. Most are bacteria, but the viruses are:

> Arboviruses, except Semlike Forest, Uganda S, Langate and yellow fever viruses
> 17D vaccine strain and Sindbis viruses

Category B2 organisms require special conditions but not special accommodation. They include:

> Hepatitis B surface antigen

Category C organisms include viruses and other microorganisms that offer no special potential hazard to laboratory workers provided the high standards of microbiological techniques and safety required by the Code are practised.

Finally, the possibilities of discovering 'new viruses' are also met by the Code. These should be treated as for B1 organisms until more information about them is available.

The virologist working on animal viral diseases that are of national importance will encounter legislation covering disease control and animal movements. Under the Diseases of Animals Act 1956 and 1975 and its associated Orders, certain viral diseases such as Aujeszkey's disease of swine, enzootic bovine leukosis, swine vesicular disease, foot-and-mouth disease, fowl pest and rabies must be notified in the event of their outbreak in the UK to the Ministry of Agriculture, Fisheries and Food (MAFF) or the local authorities, who will then take the appropriate control measures.

In addition to this, a whole barrage of legislation attempts to keep rabies at bay. The Rabies Act 1974 and Rabies (Control) Order 1974 contain similar but more stringent regulations. The quarantining of imported mammals is catered for under the Rabies (Importation of Dogs, Cats and Other Mammals) Order 1974. The Rabies Virus Order 1979 prohibits the import, keeping and using of rabies virus, except under MAFF licence or in accordance with the Medicines Act 1968. The Medicines Act 1968 operates a strict legal control over the production, supply and use of vaccines and other medicinal products. Similar legislation to that for rabies applies to a number of specific viruses mentioned under the Diseases of Animals Act.

Legislation also exists to prevent and control outbreaks of diseases of plants. The Plant Health Act 1967 and its Orders seek to control the spread of viral and other diseases of plants in the UK and include measures to prevent the introduction of diseases in imported material. The Destructive Pests and Diseases of Plants Order 1965 prohibits the landing, unloading or importing by post of any non-indigenous pest, virus or other microorganism. No specific viruses are mentioned in this legislation but it is taken that any virus that causes transmissible diseases in agriculture or horticulture or of trees or bushes is to be included.

Addresses for Further Information on Legislation

The Secretary, Dangerous Pathogens Advisory Group (Section PEH 1B), Alexander Fleming House, Elephant and Castle, London SE1 6BX.

MAFF, Animal Health Division, Hook Rise South, Tolworth, Surbiton, Surrey KT6 7NF.

MAFF, Plant Health Branch, Great Westminster House, Horseferry Road, London SW1P 2AE.

References

Cooper, M. E. (1979). The law relating to animal experiments. *Biologist* **26** (1): 33–7.

Howie, Sir J. (1978). *Code of practice for the prevention of infection in clinical laboratories and post-mortem rooms.* London: HMSO.

7.2 US LEGISLATION AND SAFETY RULES

In the USA the Department of Health, Education and Welfare has produced a list of safety rules for infectious disease laboratories entitled *Biohazard control and containment in oncogenic virus research* and edited by Hellman (1969).

Three Federal statutes — the Animal Quarantine Act of 1903, the Plant Quarantine Act of 1912 and the Federal Plant Pest Act of 1957 — prohibit the importation of pests, pathogens (including viruses) and vectors unless authorized by the Department of Agriculture. These acts and related state laws are intended to protect American agriculture by preventing the unauthorized movement and establishment of dangerous organisms. The regulations covering the shipment of pathogens and vectors of diseases of man are governed by the Department of Health, Education and Welfare of the Public Health Service.

Addresses for Further Information on Legislation

Plant Pest Control Division, United States Department of Agriculture, Federal Center Building, Hyattsville, Massachusetts 20782.

Animal Health Division, United States Department of Agriculture, Federal Center Building, Hyattsville, Massachusetts 20782.

Foreign Quarantine Program, National Communicable Disease Center, United States Public Health Service, Atlanta, Georgia 30333.

References and Further Reading

Cooper, M. E. (1979). The law relating to animal experiments. *Biologist* **26** (1) : 33–7.

Cooper, M. E. (1980). Official control affecting animals and plants used by the biologist. *Biologist* **27** (2) : 105–8.

Cooper, M. E. (1980). Biomedical law. *Biologist* **27** (4) : 183–5.

Fuscaldo, A. A., Erlick, B. J. and **Hindman, B.** (1980). *Laboratory safety: theory and practice.* London: Academic Press.

Hellmann, A. (1969). *Biohazard control and containment in oncogenic virus research.* US Department of Health, Education and Welfare.

Part II

BIBLIOGRAPHY

Bibliography

Organizations	172
Directories of organizations	172
Conferences	173
Lists of forthcoming conferences	173
Conference proceedings	173
The literature of virology	174
Journals	174
Reviews and monographic series — virology	177
Reviews and monographic series — subjects related to virology	178
Books (General; History; Bacteriophages; Chemotherapy including interferon; Immunology; Invertebrate virology; Medical virology; Methodology and diagnosis; Molecular virology; Plant virology, including fungal viruses; Slow viruses and viroids; Specific groups; Structure and morphology; Taxonomy and classification; Tumour virology; Veterinary virology, including zoonoses)	178
Encyclopedias and handbooks	188
Subject dictionaries	189
Biographical reference works	189
Searching the literature	190
Guides to the literature	190
Lists of journals	190
Abstracting and indexing journals	191
Current-awareness journals	192
Subject bibliographies (Animal viruses; Bacteriophages; Human viruses; Methodology; Plant viruses)	192
Bibliographies covering books	193

Directories of conference proceedings   194
Indexes to dissertations   194
Classification schemes   195

Miscellaneous publications mentioned in the text   195

ORGANIZATIONS

Directories of Organizations (*see* section 2.2)

1 *Directory of British associations.*   Beckenham, Kent, UK: CBD Research Ltd. 1965–.   Biennial.

2 *Directory of European associations.*   Beckenham, Kent, UK: CBD Research Ltd, 1971–.   2 vols: Part I, *Industry, Trade and Professional*; Part II, *Learned, Scientific and Technical Societies.*   Irregular.

3 *Directory of European scientific associations.*   (Edited by A. Harvey and A. Pernet.)   Harlow, Essex, UK: Longmans, 1983.   ISBN 0-582-90108-1.

4 *Encyclopedia of associations.*   Detroit: Gale Research Company, 1956–. Biennial.

5 *The grants register 1981–83.*   London: Macmillan.   ISBN 0-333-25866-5.

6 *International Association of Microbiological Societies. Directory and organization.* Washington, DC: American Society for Microbiology, 1976.

7 *International research centers directory.*   Detroit: Gale Research Company, 1981–2.   (500 pp.)   Three issues.

8 *New research centers.*   Detroit: Gale Research Company, 1965–. Quarterly.   ISSN 0028-6591.

9 *Research centers directory.*   7th edn.   Detroit: Gale Research Company, 1982.   ISSN 0080-1518.

10 *Research in British universities, polytechnics and colleges*; vol. 2, *Biological sciences — 1982.*   3rd edn.   Wetherby, West Yorkshire, UK: British Library, 1982. ISSN 0143-0734.

11 *Trade associations and professional bodies of the UK.*   Oxford: Pergamon, 1979.

12 *World guide to scientific organizations.*   3rd edn.   Detroit: Gale Research Company, 1982.

13 *World list of virus laboratories.*   5th edn.   Geneva: World Health Organization, 1979.

14 *World of learning.*   London: Europa, 1947–.   Annual.   ISSN 0084-2117.

15 *Yearbook of international organizations.* Brussels: Union of International Associations, 1949–. Biennial. ISSN 0084-3814.

CONFERENCES
Lists of Forthcoming Conferences (*see* section 3.2)

16 *Forthcoming International Scientific and Technical Conferences.* London: Aslib, 1951–. Quarterly. ISSN 0046-4686.

17 *World Meetings: outside the United States and Canada.* Chestnut Hill, New York: Macmillan, 1968–. Quarterly. ISSN 0043-8677.

18 *World Meetings: United States and Canada.* Chestnut Hill, New York: Macmillan, 1963–. Quarterly. ISSN 0043-8693.

Conference Proceedings (*see* sections 3.3, 4.4)

19 *Abstracts of Annual Meetings of the American Society for Microbiology.* Washington, DC: American Society for Microbiology, 1900–. Annual.

20 *Citrus Virus Diseases: Proceedings.* Gainesville, Florida: University Presses of Florida, for the International Organization of Citrus Virologists, 1957–. Triennial. ISSN 0074-7203.

21 *Cold Spring Harbor Conference on Cell Proliferation.* Cold Spring Harbor, New York: Cold Spring Harbor Laboratory, 1974–. Annual.

22 *Cold Spring Harbor Symposia on Quantitative Biology.* Cold Spring Harbor, New York: Cold Spring Harbor Laboratory, 1932–. Annual.

23 *European Association against Virus Diseases: Symposium.* Tampere, Finland: European Association against Virus Diseases, 1948–. Irregular.

24 *Developments in Biological Standardization.* Geneva: S. Karger, for the International Association of Biological Standardization, 1955–. Irregular.

25 *FEBS Proceedings of Meetings.* London: Academic Press, for the Federation of European Biochemical Societies, 1966–1976. Amsterdam: Elsevier/North-Holland, 1977; Oxford: Pergamon, 1978–.

26 *FEMS Symposia.* Amsterdam: Elsevier/North-Holland, for the Federation of European Microbiological Societies, 1977–. Irregular.

27 *Harvey Lectures.* London: Academic Press, for the Harvey Society of New York, 1906–. Annual.

28 *ICN–UCLA Symposium on Molecular and Cellular Biology.* London: Academic Press, 1972–. Annual.

29 *International Conference on Comparative Virology.* London: Academic Press, for the International Comparative Virology Organization, 1971–. Irregular.

30 *International Congress for Virology.* Berne: S. Karger, for the International Association of Microbiological Societies, 1968–. Irregular.

31 *Society for General Microbiology, Annual Symposium.* Cambridge: Cambridge University Press, 1951–. Annual.

THE LITERATURE OF VIROLOGY

Journals (*see* section 4.1)

32 *Acta Virologica.* London: Academic Press, for the Czechoslovak Academy of Sciences, 1957–. Bi-monthly. ISSN 0001-723X.

33 *Annals of Neurology.* Boston: Little, Brown, & Co., for the Neurological Association, 1977–. Monthly. ISSN 0364-5134.

34 *Antimicrobial Agents and Chemotherapy.* Washington, DC: American Society for Microbiology, 1972–. Monthly. ISSN 0066-4804.

35 *Antiviral Research.* Oxford: Elsevier/North-Holland Biomedical Press, 1981–. Bi-monthly. ISSN 0166-3542.

36 *Archives of Virology.* Berlin: Springer-Verlag, 1939–. Quarterly. ISSN 0304-8608.

37 *Australian Veterinary Journal.* Brunswick, Victoria: Australian Veterinary Association, 1925–. Monthly. ISSN 0005-0423.

38 *Avian Diseases.* College Station, Texas: The American Society of Avian Pathologists, 1957–. Quarterly. ISSN 0005-2086.

39 *Avian Pathology.* Huntingdon, Cambridgeshire: World Veterinary Poultry Association, 1972–. Quarterly. ISSN 0307-9457.

40 *Biochemical and Biophysical Research Communications.* New York: Academic Press, 1959–. Semi-monthly. ISSN 0006-291X.

41 *Biochimica et Biophysica Acta.* Oxford: Elsevier/North Holland Biomedical Press, 1947–. Fortnightly. ISSN 0006-3002.

42 *Cancer Research.* Washington, DC: Waverly Press, on behalf of the American Association for Cancer Research, 1941–. Monthly. ISSN 0008-5472.

43 *Cell.* Cambridge, Massachusetts: MIT Press, 1974–. Monthly Jan.–Sept., twice-monthly Oct.–Dec. ISSN 0092-8674.

44 *Comparative Immunology, Microbiology and Infectious Diseases.* Oxford: Pergamon Press, 1978–. Quarterly. ISSN 0147-9571.

45 *The Cornell Veterinarian.* Cornell University: Cornell Veterinarian Inc., 1911–. Quarterly. ISSN 0010-8901.

46 *Indian Journal of Microbiology.* Poona: Association of Microbiologists of India, 1961–. Quarterly. ISSN 0046-8991.

47 *Infection and Immunity*. Washington, DC: American Society for Microbiology, 1970–. Monthly. ISSN 0019-9567.

48 *International Journal of Cancer*. Geneva: International Union against Cancer, 1966–. Monthly. ISSN 0020-7136.

49 *Intervirology*. Basel: S. Karger, for the International Union of Microbiological Sciences, 1973–. Monthly. ISSN 0030-5526.

50 *Journal of Bacteriology*. Washington, DC: American Society for Microbiology, 1916–. Monthly. ISSN 0021-9193.

51 *Journal of Biological Chemistry*. Baltimore, Maryland: American Society of Biological Chemists, 1905–. Fortnightly. ISSN 0021-9258.

52 *Journal of Biological Standardization*. London: Academic Press, for the International Association of Biological Standardization, 1973–. Quarterly. ISSN 0092-1157.

53 *Journal of Clinical Microbiology*. Washington, DC: American Society for Microbiology, 1975–. Monthly. ISSN 0095-1137.

54 *Journal of Experimental Medicine*. New York: Rockefeller University Press, 1896–. Monthly. ISSN 0022-1007.

55 *Journal of General Microbiology*. Reading: Society for General Microbiology, 1947–. Bi-monthly. ISSN 0022-1287.

56 *Journal of General Virology*. Reading: Society for General Microbiology, 1967–. Monthly. ISSN 0022-1317.

57 *Journal of Immunology*. Baltimore, Maryland: Williams & Wilkins, 1916–. Monthly. ISSN 0022-1767.

58 *Journal of Infectious Diseases*. Chicago: Chicago University Press, 1904–. Monthly. ISSN 0022-1899.

59 *Journal of Invertebrate Pathology*. London: Academic Press, 1959–. Bi-monthly. ISSN 0022-2011.

60 *Journal of Medical Virology*. Washington, DC: Alan R. Liss Inc., 1977–. 8 issues a year. ISSN 0146-6615.

61 *Journal of Molecular Biology*. London: Academic Press, 1959–. 3 times a month. ISSN 0022-2836.

62 *Journal of the National Cancer Institute*. Washington, DC: US Government Printing Office, 1940–. Monthly. ISSN 0027-8874.

63 *Journal of Pediatrics*. St Louis: C. V. Mosby Co., 1932–. Monthly. ISSN 0022-3476.

64 *Journal of Virological Methods*. Oxford: Elsevier/North-Holland Biomedical Press, 1980–. Monthly. ISSN 0166-0934.

65 *Journal of Virology*. Washington, DC: American Society for Microbiology, 1967–. Monthly. ISSN 0022-538X.

66 *Journal of Wildlife Diseases.* Ames, Iowa: Wildlife Disease Association, 1965–. Quarterly. ISSN 0090-3558.

67 *Kitazato Archives of Experimental Medicine.* Tokyo: Kitazato Institute for Infectious Diseases, 1925–. Quarterly. ISSN 0023-1924.

68 *Lancet.* London: The Lancet Ltd, 1823–. Weekly. ISSN 0023-7507.

69 *Microbiology Letters.* Amsterdam: Elsevier/North-Holland Biomedical Press, for the Federation of European Microbiological Societies, 1977–. Monthly. ISSN 0378-1097.

70 *Microbios.* Cambridge: Faculty Press, 1969–. Three volumes a year, each volume consisting of 3 or 4 issues. ISSN 0307-5494.

71 *Molecular and Cellular Biology.* Washington, DC: American Society for Microbiology, 1981–. Monthly. ISSN 0270-7306.

72 *Molecular and General Genetics.* Berlin: Springer-Verlag, 1908–. 27 p.a. ISSN 0026-8925.

73 *National Institute of Animal Health Quarterly.* Tokyo: National Institute of Animal Health, 1961–. Quarterly. ISSN 0027-951X.

74 *Nature.* London: Macmillan Journals, 1869–. Weekly. ISSN 0028-0836.

75 *Netherlands Journal of Plant Pathology.* Wageningen: Nederlandse Planteziektenkundige Vereniging, 1895–. Bi-monthly. ISSN 0028-2944.

76 *Neurology.* New York: Modern Medicine Publications, for the National Academy of Neurology, 1951–. Monthly. ISSN 0028-3878.

77 *New England Journal of Medicine.* Boston: Massachusetts Medical Society, 1812–. Weekly. ISSN 0028-4793.

78 *Nucleic Acid Research.* London: Information Retrieval Ltd, 1974–. Bi-monthly. ISSN 0301-5610.

79 *Phytopathology.* St Paul, Minnesota: American Phytopathological Society, 1911–. Monthly. ISSN 0031-949X.

80 *Plant Disease.* St Paul, Minnesota: American Phytopathological Society, 1917–. Monthly. ISSN 0191-2917.

81 *Proceedings of the National Academy of Sciences of the United States of America.* Washington, DC: National Academy of Sciences, 1915–. Monthly. ISSN 0027-8424.

82 *Proceedings of the Royal Society of London. Series B, Biological Sciences.* London: Royal Society of London, 1832–. Irregular. ISSN 0080-4649.

83 *Proceedings of the Society for Experimental Medicine and Biology.* London: Academic Press, for the Society for Experimental Biology, 1903–. Monthly. ISSN 0037-9727.

84 *Revue Roumaine de Médecine—Virologie.* Bucharest: Editura Academie Republicii Socialiste, Romania, 1964–. Quarterly. ISSN 0035-4082.

85 *Science.* Washington, DC: American Association for the Advancement of Science, 1848–. Weekly. ISSN 0036-8075.

86 *Veterinary Microbiology.* Amsterdam: Elsevier Scientific Publishing Company, 1976–. Quarterly. ISSN 0378-1135.

87 *Veterinary Record.* London: British Veterinary Association, 1888–. Weekly. ISSN 0042-4900.

88 *Virology.* London: Academic Press, 1955–. Monthly, except semi-monthly in Jan., April, July and Oct. ISSN 0042-6822.

89 *Virus.* Tokyo: Japan Publications Trading Co., for the Society of Japanese Virologists, 1951–. Quarterly. ISSN 0042-6857.

90 *Virusy i Virusnye Zabolevaniya.* Kiev: Zdoroviia, for the Ministerostvo z Dravookhraneniia, Ukraine, USSR, 1972–. Quarterly. ISSN 0135-2083.

91 *Voprosy Virusologii.* New York: Allerton Press, for the USSR Academy of Medical Sciences and N. I. Mechnikov All-Union Scientific Society of Epidemiologists, Microbiologists and Infectionists, 1956–. Bi-monthly. ISSN 0507-4088.

92 *Zentralblatt für Bakteriologie, Mikrobiologie und Hygiene.* Berlin: Gustav Fischer Verlag, 1887–. Semi-monthly. ISSN 0174-3031.

Reviews and Monographic Series — Virology (*see* section 4.2)

93 *Advances in Virus Research.* London: Academic Press, 1953–. Annual. ISSN 0065-2172.

94 *CMI/AAB Descriptions of Plant Viruses.* Slough, Berkshire, UK: Commonwealth Agricultural Bureaux, 1966–. Annual.

95 *Comprehensive Virology.* (Edited by H. Fraenkel-Conrat and R. R. Wagner.) 18 vols. London: Plenum, 1974–1981. Irregular.

96 *Handbuch der Virusinfektionen Tieren.* Frankfurt am Main: Gustav Fischer Verlag, 1967–. Irregular.

97 *Methods in Virology.* London: Academic Press, 1967–. Irregular. ISSN 0076-6933.

98 *Monographs in Virology.* Basel: S. Karger, 1968–. Irregular. ISSN 0077-0965.

99 *NIAID Task Force Report on Virology.* Bethesda: National Institutes of Health, 1979.

100 *Perspectives in Virology.* London: Raven Press (earlier volumes by Academic Press), 1959–. Irregular. ISSN 0072-9086.

101 *Progress in Medical Virology.* Basel: S. Karger, 1958–. Annual. ISSN 0079-645X.

102 *Virology Monographs.* Berlin: Springer-Verlag, 1968–. Irregular. ISSN 0083-6591.

Reviews and Monographic Series — Subjects Related to Virology (*see* section 4.2)

103 *Advances in Cancer Research.* New York: Academic Press, 1953–. Irregular. ISSN 0065-230X.

104 *Advances in Immunology.* New York: Academic Press, 1961–. Irregular. ISSN 0065-2776.

105 *Advances in Veterinary Science and Comparative Medicine.* New York: Academic Press, 1953–. Irregular. ISSN 0065-3519.

106 *Annual Review of Genetics.* Palo Alto, California: Annual Reviews Inc., 1967–. Annual. ISSN 0066-4197.

107 *Annual Review of Microbiology.* Palo Alto, California: Annual Reviews Inc., 1947–. Annual. ISSN 0066-4227.

108 *Annual Review of Phytopathology.* Palo Alto, California: Annual Reviews Inc., 1963–. Annual. ISSN 0066-4286.

109 *Cumitechs.* Washington, DC: American Society for Microbiology, 1975–. Irregular (2–3 per year).

110 *CRC Critical Reviews in Microbiology.* Boca Raton, Florida: CRC Press, 1971–. Quarterly. ISSN 0045-6454.

111 *Current Topics in Microbiology and Immunology.* New York: Springer Verlag, 1960–. 3/4 p.a. ISSN 0070-217X.

112 *Microbiological Reviews* (formerly *Bacteriological Reviews*). Washington, DC: American Society for Microbiology, 1937–. Quarterly. ISSN 0146-0749.

113 *Microbiology (Washington).* Washington, DC: American Society for Microbiology, 1974–. Annual. ISSN 0098-1540.

114 *Progress in Nucleic Acid Research and Molecular Biology.* New York: Academic Press, 1963–. Irregular. ISSN 0079-6603.

115 *World Health Organization Monograph Series.* Geneva: World Health Organization, 1951–. Irregular. ISSN 0512-3038.

116 *World Health Organization Technical Report Series.* Geneva: World Health Organization, 1950–. Irregular.

Books (*see* section 4.3)

General

117 **Andrewes, C. H.** *The natural history of viruses.* London: Weidenfeld & Nicolson, 1967. (237 pp.)

118 **Andrewes, C. H., Pereira, H. G.** and **Wildy, P.** *Viruses of vertebrates.* 4th edn. London: Baillière Tindall, 1978. (429 pp.) ISBN 0-7020-0678-5.

119 **Bitton, G.** *Introduction to environmental virology.* Chichester, West Sussex, UK: John Wiley, 1980. (326 pp.) ISBN 0-471-04247-1.

120 **Dulbecco, R.** and **Ginsburg, H.**. *Virology.* London: Harper & Row, 1980. ISBN 0-06-140725-9.

121 **Fenner, F., McAuslan, B. R., Mims, C. A., Sambrook, J.** and **White, D. O.** *The biology of animal viruses.* 2nd edn. London: Academic Press, 1974. (834 pp.) ISBN 0-12-253040-3.

122 **Joklik, W. K.** (Ed.). *Principles of animal virology.* Hemel Hempstead, Hertfordshire, UK: Prentice-Hall, 1980. (373 pp.) ISBN 0-13-857920-5.

123 **Luria, S. E., Darnel, J. E., Baltimore, D. H. L.** and **Campbell, A.** *General virology.* 3rd edn. Chichester, West Sussex, UK: John Wiley, 1978. (660 pp.) ISBN 0-471-55640-8.

124 **Primrose, S. B.** and **Dimmock, N. J.** *Introduction to modern virology.* 2nd edn. Oxford: Blackwell Scientific, 1980. (264 pp.) ISBN 0-632-00463-0.

125 **Rhodes, A. J.** and **Van Royen, C. E.** *Textbook of virology.* 5th edn. London: Williams & Wilkins, 1968. (966 pp.)

126 **Smith, K. M.** and **Ritchie, D.** *Introduction to virology.* London: Chapman & Hall, 1980. (250 pp.) ISBN 0-412-21970-0.

History

127 **Cairns, J., Stent, G. S.** and **Watson, J. D.**. *Phage and the origins of molecular biology.* New York: Cold Spring Harbor Laboratory, 1966. (352 pp.)

128 **Hahon, N.** (Ed.). *Selected papers in virology.* Englewood Cliffs, New Jersey: Prentice-Hall, 1964. (363 pp.)

129 **Hughes, S. S.** *The virus: a history of the concept.* London: Heinemann Educational, 1977. (160 pp.) ISBN 0-434-54755-0.

130 **Reidman, S. R.** *Fighting the unseen — the story of viruses.* London: Abelard-Schuman, 1967. (144 pp.)

131 **Theiler, M.** and **Downs, W. G.** *The arthropod-borne viruses of vertebrates — an account of the Rockefeller Foundation virus program.* New Haven, Connecticut: Yale University Press, 1973. (600 pp.) ISBN 0-300-01508-9.

132 **Waterson, A. P.** and **Wilkinson, L.** *An introduction to the history of virology.* Cambridge: Cambridge University Press, 1978. (237 pp.) ISBN 0-521-21917-5.

133 **Williams, G.** *Virus hunters — the lives and triumphs of modern medical pioneers.* London: Hutchinson, 1960. (200 pp.)

Bacteriophages

134 **Adams, M. H.** *Bacteriophages.* New York: Interscience, 1959. (610 pp.)

135 **Birge, E. A.** *Bacterial and bacteriophage genetics — an introduction.* Berlin: Springer-Verlag, 1981. (359 pp.) ISBN 3-540-90504-9.

136 **Champe, S. P.** (Ed.). *Phage.* London: Academic Press, 1974. (423 pp.) ISBN 0-12-786224-2.

137 **Denhardt, D.** (Ed.). *Single-stranded DNA phages.* Cold Spring Harbor, New York: Cold Spring Harbor Laboratory, 1978. (720 pp.) ISBN 0-87969-122-0.

138 **Douglas, J.** *Bacteriophages.* London: Chapman & Hall, 1975. (136 pp.) ISBN 0-412-12640-0.

139 **Hershey, A. D.** (Ed.). *The bacteriophage lambda.* Cold Spring Harbor, New York: Cold Spring Harbor Laboratory, 1971. (792 pp.) ISBN 0-87969-102-6.

140 **Matthews, C. K.** *Bacteriophage biochemistry.* Wokingham, Berkshire, UK: Van Nostrand Reinhold, 1971. (373 pp.)

141 **Stent, G. S.** *Molecular biology of bacterial viruses.* London: W. H. Freeman, 1963. (489 pp.)

142 **Tikhonenko, A. S.** *Ultrastructure of bacterial viruses.* New York: Plenum, 1970. (294 pp.) ISBN 0-306-30421-X.

143 **Zinder, N. D.** (Ed.). *RNA phages.* Cold Spring Harbor, New York: Cold Spring Harbor Laboratory, 1975. (448 pp.) ISBN 0-87969-109-3.

Chemotherapy (Including Interferon)

144 **Bauer, D. J.** (Ed.). *Chemotherapy of viruses.* Oxford: Pergamon, 1972. (428 pp.) ISBN 0-080-16961-9.

145 **Bauer, D. J.** *Specific treatment of virus diseases.* Lancaster: MTP Press, 1977. (194 pp.) ISBN 0-852-00138-X.

146 **Gresser, I.** (Ed.). *Interferon 1 1979.* London: Academic Press, 1979. (163 pp.) ISBN 0-12-302250-9.

147 **Gresser, I.** (Ed.). *Interferon 1980*: vol. 2. London: Academic Press, 1981. (99 pp.) ISBN 0-12-302251-7.

148 **Gresser, I.** (Ed.). *Interferon 1981*: vol. 3. London: Academic Press, 1982, in press. ISBN 0-12-302252-5.

149 **Collier, L. H. and Oxford, J. S.** *Developments in antiviral therapy.* London: Academic Press, 1980. (292 pp.) ISBN 0-12-191150-6.

150 **Friedman, R. M.** *Interferons: a primer.* New York: Academic Press, 1981. (166 pp.)

151 **Geraldes, A.** *Effects of interferon on cells, viruses and the immune system.* London: Academic Press, 1975. (674 pp.) ISBN 0-12-280650-6.

152 **De Maeyer, E.** (Ed.). *The biology of the interferon system.* Amsterdam: Elsevier Biomedical Press, 1981. (488 pp.) ISBN 0-686-80609-3.

153 **Oxford, J. S., Drasar, F. A.** and **Williams, J. D.** (Eds). *Chemotherapy of herpes simplex virus infection.* London: Academic Press, 1977. (154 pp.) ISBN 0-12-531760-3.

154 **Oxford, J. S.** and **Williams, J. D.** (Eds). *Chemotherapy and control of influenza.* London: Academic Press, 1976. (95 pp.) ISBN 0-12-531750-6.

155 **Smith, R. A.** and **Kirkpatrick, W.** (Eds). *Ribavirin — a broad spectrum antiviral agent.* London: Academic Press, 1980. (265 pp.) ISBN 0-12-652350-9.

Immunology

156 **Duffy, J. I.** (Ed.). *Vaccine preparation techniques.* Park Ridge, NJ: Noyes Data Corporation, 1980. (403 pp.) ISBN 0-8155-0796-8.

157 **Koprowski, C.** and **Koprowski, H.** (Eds). *Viruses and immunity.* London: Academic Press, 1975. (156 pp.) ISBN 0-12-420350-7.

158 **Kurstak, E.** and **Morisset, R.** (Eds.) *Viral immunodiagnosis.* London: Academic Press, 1974. (352 pp.) ISSN 0-12-429750-1.

159 **Notkins, A. L.** (Ed.). *Viral immunology and immunopathology.* London: Academic Press, 1975. (498 pp.) ISBN 0-12-522050-2.

160 **Roitt, I. M.** *Essential immunology.* 4th edn. Oxford: Blackwell Scientific, 1980. (358 pp.) ISBN 0-632-00739-7.

161 **Rose, N. R.** and **Friedman, H.** (Eds). *Manual of clinical immunology.* 2nd edn. Washington, DC: American Society for Microbiology, 1980. (1105 pp.) ISBN 0-914826-25-5.

162 **Rosenburg, N.** and **Cooper, L.** *Vaccines and viruses.* New York: Grosset & Dunlap, 1971. (159 pp.) ISBN 0-448-21414-8.

Invertebrate Virology

163 **Gibbs, A. J.** (Ed). *Viruses and invertebrates.* Amsterdam: North-Holland, 1973. (690 pp.) ISBN 0-444-10529-8.

164 **Harris, K. F.** and **Maramorosch, K.** (Eds). *Aphids as virus vectors.* London: Academic Press, 1977. (559 pp.) ISBN 0-12-327550-4.

165 **Smith, K. M.** *Insect virology.* London: Academic Press, 1967. (268 pp.)

166 **Smith, K. M.** *Virus–insect relationships.* Harlow, Essex, UK: Longmans, 1976. (308 pp.) ISBN 0-582-46612-1.

Medical Virology

167 **Acton, J. D.** *Fundamentals of medical virology for students of medicine and related sciences.* Philadelphia: Lea & Febiger, 1974. (358 pp.) ISBN 0-8121-0433-1.

168 **Davis, B. D.** *Microbiology.* 3rd edn. London: Harper & Row, 1980. (1274 pp.) ISBN 0-06-140691-0.

169 **Duguid, J. P., Marmion, B. P.** and **Swain, R. H. A.** (Eds). *Medical microbiology.* 13th edn. Edinburgh: Churchill Livingstone, 1978. (666 pp.) ISBN 0-443-01789-1.

170 **Evan, A. S.** (Ed.). *Viral infections of humans: epidemiology and control.* Chichester, West Sussex, UK: John Wiley, 1976. (584 pp.) ISBN 0-471-99435-9.

171 **Fenner, F. J.** and **White, D. O.** *Medical virology.* 2nd edn. London: Academic Press, 1976. (520 pp.) ISBN 0-12-25306-0.

172 **Hooks, J. J.** and **Gordon, G. W.** *Viral infections in oral medicine.* Amsterdam: Elsevier Biomedical Press, 1982. (352 pp.) ISBN 0-444-00674-5.

173 **Kurstak, E.** and **Kurstak, C.** *Human and related viruses.* London: Academic Press, 1977. 2 vols (555 pp. each). ISBN 0-12-429701-3.

174 **Pumper, R. W.** and **Yamashiroya, H. M.** *Essentials of medical virology.* Eastbourne, East Sussex, UK: Holt-Saunders, 1975. (250 pp.) ISBN 0-7216-7406-2.

175 **Rothschild, H.** (Ed.). *Human diseases caused by viruses.* Oxford: Oxford University Press, 1978. (361 pp.) ISBN 0-19-502286-6.

176 **Timbury, M. G.** *Notes on medical virology.* Edinburgh: Churchill Livingstone, 1974. (161 pp). ISBN 0-443-01100-1.

177 **Waterson, A. P.** (Ed.). *Recent advances in clinical virology.* Edinburgh: Churchill Livingstone, 1977. (200 pp.) ISBN 0-443-01542-2.

Methodology and Diagnosis

178 **Agricultural Research Council.** *Methods for the detection of the viruses of certain diseases in animals and animal products.* Luxembourg: Office of Official Publications of the European Communities, 1976.

179 **Ball, E.** *Serological tests for the identification of plant viruses.* St Paul, Minnesota: American Phytopathological Society, 1974. (32 pp.)

180 **Christensen, M.** *Basic laboratory procedures in diagnostic virology.* Springfield, Illinois: Charles C. Thomas, 1977. (115 pp.) ISBN 0-398-03617-9.

181 **Gardner, P. S.** and **McQuillin, J.** *Rapid virus diagnosis.* 2nd edn. London: Butterworths, 1980. (317 pp.) ISBN 0-407-38441-3.

182 **Grist, N. R.** *et al.* *Diagnostic methods in clinical virology.* 3rd edn. Oxford: Blackwell Scientific, 1979. (240 pp.) ISBN 0-632-00152-6.

183 **Howard, C. R.** (Ed.). *New developments in practical virology.* New York: Liss, 1982. (343 pp.) ISBN 0-8451-1654-1.

184 **Kuchler, R. J.** *Biochemical methods in cell culture and virology.* New York: Academic Press, 1977. (352 pp.) ISBN 0-12-786880-1.

185 **Kurstak, E.** and **Kurstak, C.** (Eds). *Comparative diagnosis of viral diseases.* London: Academic Press, 1978–1981. 4 vols: vol. 1 (1978, 555 pp.), ISBN 0-12-429701-3; vol. 2 (1979, 555 pp.), ISBN 0-12-429702-1; vol. 3

(1981, 429 pp.), ISBN 0-12-429703-X; vol. 4 (1981, 716 pp.), ISBN 0-12-429704-8.

186 **Lennette, D. A.** and **Lennette, E. H.** *A user's guide to the diagnostic virology laboratory.* Baltimore, Maryland: University Park Press, 1980. (103 pp.) ISBN 0-8391-1623-3.

187 **Lennette, E. H.** (Ed.). *Manual of clinical microbiology.* 3rd edn. Washington, DC: American Society for Microbiology, 1980. (1044 pp.) ISBN 0-914826-24-7.

188 **Lennette, E. H.** and **Schmidt, N. J.** (Eds). *Diagnostic procedures for viral, rickettsial and chlamydial infections.* 5th edn. Washington, DC: American Public Health Association, 1979. (1138 pp.) ISBN 0-87553-087-7.

189 **Maniatis, T., Fritsch, E. F.** and **Sambrook, J.** *Molecular cloning.* Cold Spring Harbor, New York: Cold Spring Harbor Laboratory, 1981. (400 pp.) ISBN 0-87969-136-0.

190 **Paul, J.** *Cell and tissue culture.* 5th edn. Edinburgh: Churchill Livingstone, 1975. (488 pp.) ISBN 0-443-01298-9.

191 **Rovozzo, G. C.** and **Burke, C. N.** *A manual of basic virological techniques.* Hemel Hempstead, Hertfordshire, UK: Prentice-Hall, 1973. (287 pp.) ISBN 0-13-553289-2.

192 **Seligson, D.** *Handbook series in clinical laboratory science. Section H, Virology and rickettsiology.* West Palm Beach, Florida: CRC Press, 1979. 2 vols (488 pp. each.) ISBN 0-8493-7061-2.

193 **Valle, M.** *Factors affecting plaque assay of animal viruses.* Copenhagen: Munksgaard, 1971. (69 pp.) ISBN 87-16-00696-8.

Molecular Virology

194 [**Anon.**]. *Studies on viral replication.* Luxembourg: Commission of the European Communities, 1976. (424 pp.)

195 **Barry, R. D.** and **Mahy, B. W. J.** *The biology of large RNA viruses.* London: Academic Press, 1970. (736 pp.) ISBN 0-12-07950-X.

196 **Blough, H. A.** and **Tiffany, J. M.** *Cell membranes and viral envelopes.* London: Academic Press, 1980. 2 vols: vol. 1 (457 pp.), ISBN 0-12-107201-0; vol. 2 (389 pp.), ISBN 0-12-107202-9.

197 **Hayes, W.** *The genetics of bacteria and their viruses.* Oxford: Blackwell Scientific, 1976. (177 pp.) ISBN 0-632-00488-6.

198 **Kaper, J. M.** *The chemical basis of virus structure dissociation and reassembly.* Amsterdam: North-Holland, 1975. (485 pp.) ISBN 0-7204-7139-7.

199 **Knight, C. A.** *Chemistry of viruses.* 2nd edn. Berlin: Springer-Verlag, 1975. (335 pp.) ISBN 0-387-06772-8.

200 **Knight, C. A.** *Molecular virology.* Maidenhead, Berkshire, UK: McGraw-Hill, 1974. (235 pp.) ISBN 0-07-035112-0.

201 **Koch, G.** and **Richter, D**. (Eds). *Biosynthesis, modification and processing of cellular and viral polyprotein.* London: Academic Press, 1980. (341 pp.) ISBN 0-12-417560-0.

202 **Levy, H. B.** (Ed.). *The biochemistry of viruses.* New York: Marcel Dekker, 1969. (657 pp.)

203 **Lonberg-Holm, K.** and **Philipson, L.** *Virus receptors.* Part II, *Animal viruses.* London: Chapman & Hall, 1980. (216 pp.) ISBN 0-412-16410-8.

204 **Martin, S. J.** *The biochemistry of viruses.* Cambridge: Cambridge University Press, 1978. (168 pp.) ISBN 0-521-21678-8.

205 **Pennington, H.** and **Ritchie, D. A.** *Molecular virology.* London: Chapman & Hall, 1975. (64 pp.) ISBN 0-412-12590-0.

206 **Prosad-Nayak, D.** *The molecular biology of animal viruses.* Basel: Marcel Dekker, 1977. 2 vols: vol. 1 (552 pp.), ISBN 0-8247-6533-8; vol. 2 (430 pp.), ISBN 0-8247-6534-6.

207 **Randall, L. L.** and **Philipson, L.** *Virus receptors.* Part I, *Bacterial viruses.* London: Chapman & Hall, 1980. (160 pp.) ISBN 0-412-15660-1.

208 **Wolstenholme, G. E. W.** and **O'Connor, M.** *Strategy of the viral genome.* Edinburgh: Churchill Livingstone, 1971. (406 pp.)

Plant Virology (Including Fungal Viruses)

209 **Cooper, J. I.** *Virus diseases of trees and shrubs* Oxford: Institute of Terrestrial Ecology, 1979. (74 pp.) ISBN 0-904282-28-7.

210 **Gibbs, A.** and **Harrison, B.** *Plant virology. The principles.* London: Edward Arnold, 1979. (302 pp.) ISBN 0-7131-2764-3.

211 **King, R. C.** (Ed.). *Plants, plant viruses and protists.* New York: Plenum, 1974. (631 pp.) ISBN 0-306-37612-1.

212 **Kurstak, E.** *Handbook of plant virus infections. Comparative diagnosis.* Amsterdam: Elsevier/North-Holland Biomedical Press, 1981. (944 pp.) ISBN 0-444-80309-2.

213 **Lemka, P. A.** (Ed.). *Viruses and plasmids in fungi.* Basel: Marcel Dekker, 1979. (664 pp.) ISBN 0-8247-6916-3.

214 **Matthews, R. E. F.** *Plant virology.* London: Academic Press, 1976. (778 pp.) ISBN 0-12-480560-4.

215 **Molitoris, H. P.** (Ed.). *Proc. 12th International Congress for Microbiology, Munich, 1978: Fungal viruses.* Berlin: Springer-Verlag, 1979. (194 pp.) ISBN 0-387-09477-6.

216 **Smith, K. M.** (Ed.). *Textbook of plant virus diseases.* 3rd edn. London: Academic Press, 1972. (648 pp.)

217 **Smith, K. M.** *Plant viruses.* London: Chapman & Hall, 1977. (248 pp.) ISBN 0-412-14710-6.

218 **Taylor, C. E.** and **Seinhorst, J. W.** *Nematode vectors of plant viruses.* New York: Plenum Press, 1975. (470 pp.) ISBN 0-306-35602-2.

Slow Viruses and Viroids

219 **Adams, D. H.** and **Bell, T. M.** *Slow viruses.* London: Addison-Wesley Advanced Book Program. 1976. (224 pp.) ISBN 0-201-00042-3.

220 **Diener, D. O.** *Viroids and viroid diseases.* Chichester, West Sussex, UK: John Wiley, 1979. (270 pp.) ISBN 0-471-103504-1.

221 **Kimberlin, R. H.** *Slow virus diseases of animals and man.* Amsterdam: North Holland, 1976. (404 pp.) ISBN 0-7204-0418-5.

222 **Prusiner, S. B.** and **Hadlow, W. J.** (Eds). *Slow transmissible diseases of the nervous system.* London: Academic Press, 1979. 2 vols: vol. 1 (496 pp.), ISBN 0-12-566301-3; vol. 2 (526 pp.), ISBN 0-12-566302-1.

Specific Groups

223 **Baer, G. M.** (Ed.). *The natural history of rabies.* London: Academic Press, 1975. 2 vols: vol. 1 (454 pp.), ISBN 0-12-072401-4; vol. 2 (387 pp.), ISBN 0-12-072402-2.

224 **Bishop, D. H. L.** (Ed.). *Rhabdoviruses.* West Palm Beach, Florida: CRC Press, 1981. 3 vols: (approx. 200 pp. each). Vol. 1, ISBN 0-8493-5913-9; vol. 2, ISBN 0-8493-5914-7; vol. 3, ISBN 0-8493-5915-5.

225 **Bricout, E.** and **Scherrer, R.** (Eds.). *Viral enteritis in humans and animals.* Paris: INSERM, 1980. (450 pp.) ISBN 2-85598-193-X.

226 **Fraser, K. B.** and **Martin, S. J.** *Measles virus and its biology.* London: Academic Press, 1978. (262 pp.) ISBN 0-12-265350-5.

227 **Horzinek, M. C.** *Non-arthropod-borne togaviruses.* London: Academic Press, 1981. (200 pp.) ISBN 0-12-356550-2.

228 **Kaplan, A. S.** (Ed.). *The herpesviruses.* London: Academic Press, 1974. (768 pp.) ISBN 0-12-397050-4.

229 **Kaplan, C.** *Rabies — the facts.* Oxford: Oxford University Press, 1977. (124 pp.) ISBN 0-19-264918-3.

230 **Kilbourne, E. D.** (Ed.). *The influenza viruses and influenza.* London: Academic Press, 1975. (573 pp.) ISBN 0-12-407050-7.

231 **Kurstak, E.** (Ed.). *Arctic and tropical arboviruses.* London: Academic Press, 1979. (352 pp.) ISBN 0-12-429765-X.

232 **Mahy, B. W. J.** and **Barry, R. D.** (Eds). *Negative strand viruses.* London: Academic Press, 1973. 2 vols: vol. 1 (862 pp.), ISBN 0-12-465301-4; vol. 2 (475 pp.), ISBN 0-12-465302-2.

233 **Mahy, B. W. J.** and **Barry, R. D.** (Eds.). *Negative strand viruses and the host cell.* London: Academic Press, 1977. (886 pp.) ISBN 0-12-465350-2.

234 **Nahmias, A. J., Dowdle, W. R.** and **Schinazi, R. F.** (Eds). *The human herpesviruses.* Amsterdam: Elsevier Biomedical Press, 1981. (736 pp.) ISBN 0-444-00553-6.

235 **Perez-Bercoff, R.** (Ed.). *The molecular biology of picornaviruses.* London: Plenum, 1979. (371 pp.) ISBN 0-306-40192-4.

236 **Schlesinger, R. W.** (Ed.). *The togaviruses: biology, structure and replication.* London: Academic Press, 1980. (720 pp.) ISBN 0-12-625380-3.

237 **Tyrell, D. A. J.** and **Pereira, H. G.** (Eds). *Influenza. A Royal Society discussion.* London: Royal Society, 1980. (172 pp.) ISBN 0-85403-138-3.

238 **Ward, D. C.** and **Tattersal, P.** (Eds). *Replication of mammalian parvoviruses.* Cold Spring Harbor, New York: Cold Spring Harbor Laboratory, 1978. (547 pp.) ISBN 0-87969-120-4.

239 **Zuckerman, A. J.** and **Howard, C. R.** *Hepatitis viruses of man.* London: Academic Press, 1980. (278 pp.) ISBN 0-12-782150-3.

Structure and Morphology

240 **Dalton, A. J.** and **Haguenau, F.** (Eds). *Ultrastructure of animal viruses and bacteriophages — an atlas.* London: Academic Press, 1973. (430 pp.) ISBN 0-12-200960-6.

241 **Horne, R. W.** *Virus structure.* London: Academic Press, 1974. (54 pp.) ISBN 0-12-355750-X.

242 **Horne, R. W.** *Structure and function of viruses.* London: Edward Arnold, 1978. (54 pp.) ISBN 0-7131-2706-6.

243 **Madely, C. R.** *Virus morphology.* Edinburgh: Churchill Livingstone, 1972. (185 pp.)

244 **Maramorosch, K.** (Ed.). *An atlas of insect and plant viruses.* London: Academic Press, 1977. (478 pp.) ISBN 0-12-470275-9.

245 **Williams, R. C.** and **Fisher, H. W.** *An electron microscopic atlas of viruses.* Springfield, Illinois: Charles C. Thomas, 1974. (151 pp.) ISBN 0-398-03153-3.

Taxonomy and Classification

246 **Hanson, H. P.** (Ed.). *Contributions to the systematic plant virology. II. Codes, data and taxonomy.* Copenhagen: DSR Forlag, 1975. (25 pp.)

247 **Matthews, R. E. F.** *Classification and nomenclature of viruses.* Basel: S. Karger, 1979. (296 pp.) ISBN 3-8055-0523-X.

248 **Matthews, R. E. F.** (Ed.). Classification and nomenclature of viruses. 4th report. *Intervirology* **17** (1–3), 4–199 (1982).

Tumour Virology

249 **Bachmann, P. A.** (Ed.). *Leukaemias, lymphomas and papillomas: comparative aspects (Proceedings of the 5th Munich Symposium on Microbiology, 1980).* London: Taylor & Francis, 1980. (273 pp.) ISBN 0-85066-213-3.

250 **Ceglowski, W. S.** and **Friedman, H.** (Eds). *Virus tumorigenesis and immunogenesis.* London: Academic Press, 1973. (424 pp.) ISBN 0-12-165050-2.

251 **Crowell, R., Friedman, H.** and **Praer, J. E.** (Eds). *Tumor virus infections and immunity.* Baltimore, Maryland: University Park Press, 1976. (371 pp.) ISBN 0-8391-0891-5.

252 **Dulbecco, R.** *The biology of small DNA tumor viruses.* New York: MSS Information Corporation, 1974. ISBN 0-8422-7258-5.

253 **Giraldo, G.** and **Beth, E.** *The role of viruses in human cancer,* vol. 1. Amsterdam: Elsevier/North-Holland, 1980. (352 pp.) ISBN 0-444-00440-8.

254 **Gross, L.** *Oncogenic viruses.* 2nd edn. Oxford: Pergamon, 1970. (994 pp.) ISBN 0-080-13236-7.

255 **Ikawa, Y.** and **Odaka, T.** *Oncogenic viruses and host cell genes.* London: Academic Press, 1979. (491 pp.) ISBN 0-12-370650-5.

256 **Kaplan, A. S.** (Ed.). *Viral transformation and endogenous viruses.* London: Academic Press, 1975. (184 pp.) ISBN 0-12-397060-1.

257 **Klein, G.** (Ed.). *Viral oncology.* New York: Raven Press, 1980. (860 pp.) ISBN 0-89004-390-6.

258 **Nicolau, C.** (Ed.). *Virus-transformed cell membranes.* London: Academic Press, 1978. (421 pp.) ISBN 0-12-518650-9.

259 **Schiminovich, S.** *The biology of DNA tumor viruses.* London: Plenum Press, 1976. (273 pp.) ISBN 0-306-32201-3.

260 **Stephenson, J. R.** (Ed.). *Molecular biology of RNA tumor viruses.* London: Academic Press, 1980. (527 pp.) ISBN 0-12-66050-6.

261 **Temin, H. M.** *The biology of RNA tumor viruses.* New York: MSS Information Corporation, 1974. (318 pp.) ISBN 0-8422-7234-8.

262 **de Thé, G., Henle, W.** and **Rapp, F.** (Eds). *Oncogenesis and the herpesviruses III.* Lyon: International Agency for Research in Cancer, 1978. 2 vols: Part I (580 pp.); Part II (520 pp.). ISBN 92-832-1124-3.

263 **Todaro, G. J.** *Investigation of oncogenic viruses.* New York: MSS Information Corporation, 1974. Vol. 1 (234 pp.), ISBN 0-8422-7234-7. Vol. 2 was never published.

264 **Tooze, J.** (Ed.). *DNA tumor viruses.* 2nd edn. Cold Spring Harbor, New York: Cold Spring Harbor Laboratory, 1981. (1100 pp.) ISBN 0-87969-141-7.

265 **Tooze, J.** (Ed.). *Selected papers in tumor virology.* Cold Spring Harbor, New York: Cold Spring Harbor Laboratory, 1974. (1040 pp.) ISBN 0-87969-112-3.

266 **Weiss, R. A., Teich, N. M., Varmus, H. E.** and **Coffin, J. M.** (Eds). Cold Spring Harbor, New York: Cold Spring Harbor Laboratory, 1982. (1396 pp.) ISBN 0-87969-132-8.

267 **Yohei, I.** (Ed.). *Viruses and human cancer.* Basel: S. Karger, 1978. (306 pp.) ISBN 3-8055-2701-2.

Veterinary Virology (Including Zoonoses)

268 **Beron, G. W.** (Ed.). *Viral zoonoses.* West Palm Beach, Florida: CRC Press, 1981. 2 vols (460 pp. each). ISBN 0-8493-2911-6 and 0-8493-2912-4.

269 **Buxton, A.** and **Fraser, G.** *Animal microbiology*; vol. 2, *Rickettsias and viruses.* Oxford: Blackwell Scientific, 1977. (600 pp.) ISBN 0-632-0094-1.

270 **Gibbs, E. P. J.** (Ed.). *Virus diseases of food animals;* vol. 1, *International perspectives.* London: Academic Press, 1981. (350 pp.) ISBN 0-12-282201-3.

271 **Gibbs, E. P. J.** (Ed.). *Virus diseases of food animals;* vol. 2, *Disease monographs.* London: Academic Press, 1981. (450 pp.) ISBN 0-12-282202-1.

272 **Gillespie, J. H.** and **Timoney, J. F.** *Hagan and Bruner's Infectious diseases of domestic animals.* 7th edn. Ithaca, New York: Cornell University Press, 1981. (851 pp.) ISBN 0-8014-1333-8.

273 **Hofstad, M. S., Calnek, B. W., Helmboldt, C. F., Reid, W. M.** and **Yoder, H. W.** *Diseases of poultry.* 7th edn. Ames, Iowa: Iowa State University Press, 1978. (960 pp.) ISBN 0-8138-0430-2.

274 **Kahrs, R. F.** *Viral diseases of cattle.* Ames, Iowa: Iowa State University Press, 1981. (300 pp.) ISBN 0-8138-0860-X.

275 **Mohanty, S. B.** and **Dutta, S. K.** *Veterinary virology.* Philadelphia, Pennsylvania: Lea & Febiger, 1981. (372 pp.) ISBN 0-8121-0777-2.

276 **Postgraduate Committee in Veterinary Science.** *Advances in veterinary virology.* Sydney: University of Sydney, 1982. (541 pp.)

Encyclopedias and Handbooks (see section 4.6)

277 *Biology data book.* 2nd edn. Bethesda, Maryland: Federation of American Societies for Experimental Biology, 1974. ISBN 0-913822-08-6.

278 *Cell biology.* Bethesda, Maryland: Federation of American Societies for Experimental Biology, 1976. ISBN 0-913822-10-8.

279 *Encyclopaedia Britannica.* Chicago: Encyclopaedia Britannica, 1975. ISBN 0-85229-297-X.

280 *Handbook of microbiology.* (Edited by A. I. Laskin and H. A. Lechavilier.) 4 vols. Boca Raton, Florida: Chemical Rubber Company, 1973. ISBN 0-87819-580-7.

281 *McGraw-Hill encyclopedia of science and technology.* 5th edn. New York: McGraw-Hill, 1982.

282 *Van Nostrand's scientific encyclopedia.* 6th edn. (Edited by D. M. Considine.) (3100 pp.) New York: Van Nostrand, 1983.

Subject Dictionaries (*see* section 4.6)

283 **Cowan, S. T.** and **Hill, L. R.** (Eds). *A dictionary of microbial taxonomy.* Cambridge: Cambridge University Press, 1978. (285 pp.) ISBN 0-521-21390-X.

284 **Dupayrat, J.** *A dictionary of biomedical acronyms.* Paris: SCM, 1978. (109 pp.) ISBN 2-901133-07-X.

285 **Evans, A.** *Glossary of molecular biology.* London: Butterworths, 1974. (55 pp.) ISBN 0-408-70640-6.

286 **Herbot, W. J.** and **Wilkinson, P. C.** (Eds). *A dictionary of immunology.* 2nd edn. Oxford: Blackwell Scientific, 1977. (192 pp.) ISBN 0-632-00135-6.

287 **Rieger, R., Michaelis, A.** and **Green, M. M.** *Glossary of genetics and cytogenetics.* New York: Springer-Verlag, 1976. (647 pp.) ISBN 0-387-07668-9.

288 **Rowson, K. E. K., Rees, T. A. L.** and **Mahy, B. W. J.** *A dictionary of virology.* Oxford: Blackwell Scientific, 1981. (230 pp.) ISBN 0-632-0069-7.

289 **Singleton, P.** and **Sainsbury, D.** *Dictionary of microbiology.* Chichester, West Sussex, UK; Wiley Interscience, 1981. (490 pp.) ISBN 0-471-99658-0.

290 **Steneish, J.** *A dictionary of biochemistry.* Chichester, West Sussex, UK: Wiley Interscience, 1975. (344 pp.) ISBN 0-471-82105-5.

Biographical Reference Works (*see* section 4.6)

291 *American men and women of science; physical and biological sciences.* New York: Jaques Cattell Press (Bowker), 1906–. Irregular. ISSN 0065-9347.

292 *Biographical memoirs of the National Academy of Sciences.* Washington, DC: National Academy of Sciences, 1953–. Irregular.

293 *Biographical Memoirs of Fellows of the Royal Society.* London: Royal Society of London, 1955–. Annual. ISSN 0080-4606.

294 **Waterson, A. P.** and **Wilkinson, L.** *An introduction to the history of virology.* Cambridge: Cambridge University Press, 1978. (237 pp.) ISBN 0-521-21917-5.

295 *Who was who.* London: A & C Black, 1967–1981.
Vol. 1   1897–1915   ISBN 0-7136-0168-X.
Vol. 2   1916–1928   ISBN 0-7136-0169-8.
Vol. 3   1929–1940   ISBN 0-7136-0170-1.
Vol. 4   1941–1950   ISBN 0-7136-0171-X.

Vol. 5   1951–1960   ISBN 0-7136-0172-8.
Vol. 6   1961–1970   ISBN 0-7136-2008-0.
Vol. 7   1971–1980   ISBN 0-7136-2176-1.
also index   1897–1980   ISBN 0-7136-2177-X.

296 *Who's who.* London: A. & C. Black, 1849–. Annual. ISSN 0083-937X.

297 *Who's who in British scientists 1980/81.* 3rd edn. London: Simon Books, 1980. ISBN 0-86229-001-5.

SEARCHING THE LITERATURE

Guides to the Literature (*see* section 5.1)

298 **Bottle, R. T.** and **Wyatt, H. V.** (Eds). *The use of biological literature.* 2nd edn. London: Butterworths, 1971. (392 pp.) ISBN 0-408-01221-4. [A new edition is in preparation.]

299 *Directory of cancer research information resources.* 2nd edn. Bethesda, Maryland: US National Cancer Institute, 1979. (250 pp.)

300 **Kerker, A. E.** and **Murphey, H. T.** *Biological and biomedical resources literature.* Lafayette, Indiana: Purdue University, 1968. (300 pp.)

301 **Kerker, A. E.** and **Murphey, H. T.** *Comparative and veterinary medicine: a guide to the resource literature.* Madison: University of Wisconsin Press, 1973. (308 pp.) ISBN 0-299-06330-5.

302 **Lilley, G. P.** (Ed.). *Information sources in agriculture and food science.* London: Butterworths, 1981. (603 pp.) ISBN 0-408-10612-3.

303 **Morton, L. T.** (Ed.). *Use of medical literature.* London: Butterworths, 1977. (462 pp.) ISBN 0-408-70916-2.

304 **Smith, R. C.** and **Reid, W. M.** *Guide to the literature of the life sciences.* 8th edn. Minneapolis: Burgess, 1972. (166 pp.) ISBN 0-8087-1964-5.

Lists of Journals (*see* section 5.2)

305 *British Union Catalogue of Periodicals (BUCOP).* London: Butterworths, 1955–1981. Quarterly.

306 *Current serials received.* Wetherby, West Yorkshire, UK: British Library Lending Division, 1965–. Annual. ISSN 0309-0655.

307 *Irregular serials and annuals.* New York: Bowker, 1967–. Biennial. ISSN 0000-0043.

308 *Keyword Index to Serial Titles.* Wetherby, West Yorkshire, UK: British Library Lending Division, 1979–. Updated quarterly.

309 *New Serial Titles.* Washington, DC: Library of Congress, 1953–. Monthly. ISSN 0028-6680.

310 *New Serial Titles — Classed Subject Arrangement.* Washington, DC: Library of Congress, 1955–. Monthly. ISSN 0028-6699.

311 *Periodicals relevant to microbiology and immunology.* Marseille: International Union of Microbiological Societies, 1968.

312 *Serials in the British Library.* London: British Library, 1981–. Quarterly. ISSN 0260-0005.

313 *Ulrich's International periodical directory.* New York: Bowker, 1932–. Annual. ISSN 0000-0175.

314 *Ulrich's Quarterly.* New York: Bowker, 1977–. Quarterly. ISSN 0000-0507.

Abstracting and Indexing Journals (*see* section 5.3)

315 *Animal Disease Occurrence.* Slough, Berkshire, UK: Commonwealth Agricultural Bureaux, 1980–. Bi-annual. ISSN 0144-3879.

316 *Biological Abstracts.* Philadelphia: Bioscience Information Service, 1927–. Semi-monthly. ISSN 0192-6935.

317 *Bulletin Signalétique.* Part 340: *Microbiologie—Virologie—Immunologie.* Paris: Centre National de la Recherche Scientifique, 1961–. Monthly. ISSN 0007-5450.

318 *Excerpta Medica.* Amsterdam: Excerpta Medica, 1971–. Monthly. ISSN 0031-6520.

319 *Foot-and-Mouth Disease Bulletin.* Pirbright, Surrey, UK: Wellcome Research Laboratories, 1962–. Monthly.

320 *ICRDB Cancergrams.* Bethesda, Maryland: US National Cancer Institute, 1977–. Monthly.

321 *Index Medicus.* Bethesda, Maryland: National Library of Medicine, 1960–. Monthly. ISSN 0019-3879.

322 *Index Veterinarius.* Slough, Berkshire, UK: Commonwealth Agricultural Bureaux, 1933–. Monthly. ISSN 0019-4123.

323 *International Abstracts of Biological Sciences.* Oxford: Pergamon Press, 1956–. Monthly. ISSN 0020-5818.

324 *Review of Plant Pathology.* Slough, Berkshire, UK: Commonwealth Agricultural Bureaux, 1922–. Monthly. ISSN 0034-6438.

325 *Science Citation Index.* Philadelphia: Institute for Scientific Information, 1961–. Monthly. ISSN 0036-827X.

326 *Veterinary Bulletin.* Slough, Berkshire, UK: Commonwealth Agricultural Bureaux, 1931–. Monthly. ISSN 0042-4854.

327 *Virology Abstracts.* London: Information Retrieval Ltd, 1967–. Monthly. ISSN 0042-6830.

328 *Virology Literature.* [s.l.] : [s.n.], 1973.

329 *Zentralblatt für Bakteriologie, Parasitenkunde, Infektionskrankheiten und Hygiene.* Berne: Gustav Fischer Verlag, 1879–. Monthly. ISSN 0044-4073.

Current-Awareness Journals (*see* section 5.3)

330 *Current Advances in Plant Science.* Oxford: Pergamon Press, 1972–. Monthly. ISSN 0044-4073.

331 *Current Contents — Life Sciences.* Philadelphia: Institute for Scientific Information, 1958–. Weekly. ISSN 0011-3409.

332 *Current Contents — Agriculture, Biology and Environmental Sciences.* Philadelphia: Institute for Scientific Information, 1970–. Weekly. ISSN 0090-0508.

Subject Bibliographies (*see* section 5.4)

Animal Viruses

The following were all compiled by the Comonwealth Bureau of Animal Health.

333 *Bovine leukosis 1972–1978.* Slough, Berkshire, UK: Commonwealth Agricultural Bureaux, 1978. (192 pp.)

334 *Bursa of Fabricius in health and disease and infectious bursal (gumboro) disease 1975–1978.* Slough, Berkshire, UK: Commonwealth Agricultural Bureaux, 1978. (36 pp.)

335 *Rabies: epidemiology, pathology, immunology, diagnosis and control 1976–1978.* Slough, Berkshire, UK: Commonwealth Agricultural Bureaux, 1978. (27 pp.)

The following were all compiled by the Emergency Programs Foreign Animal Disease Data Bank.

336 *Foot and mouth.* Hyattsville, Maryland: US Department of Agriculture, 1980. (443 pp.)

337 *Venezuelan equine encephalomyelitis.* Hyattsville, Maryland: US Department of Agriculture, 1980. (93 pp.)

338 *Hog Cholera.* Hyattsville, Maryland: US Department of Agriculture, 1981. (209 pp.)

339 *Newcastle disease.* Hyattsville, Maryland: US Department of Agriculture, 1978. (233 pp.)

340 *Ephemeral fever.* Hyattsville, Maryland: US Department of Agriculture, 1981. (20 pp.)

341 *Visna maedi.* Hyattsville, Maryland: US Department of Agriculture, 1980. (52 pp.)

342 *Swine Vesicular Disease.* Hyattsville, Maryland: US Department of Agriculture, 1980. (33 pp.)

343 **Ministry of Agriculture, Fisheries and Food.** *Rabies and its control: a selected bibliography of English language publications.* Tolworth, Surrey, UK: Ministry of Agriculture, Fisheries and Food, 1978. (9 pp.)

344 **Ministry of Agriculture, Fisheries and Food.** *Foot-and-mouth disease and its implications for Great Britain.* Tolworth, Surrey, UK: Ministry of Agriculture, Fisheries and Food, 1979. (27 pp.)

Bacteriophages

345 **Raettig, H.** *Bakteriophagie.* Stuttgart: Gustav Fischer Verlag. 2 vols: vol. 1 (1958, 588 pp.); vol. 2 (1965, 2 parts, 750 pp.)

Human Viruses

346 **The Biological Carcinogenesis Branch.** *Biological carcinogenesis research bibliography.* Bethesda, Maryland: US National Cancer Institute, 1965–.

347 **National Library of Medicine.** *Index Medicus.* Bethesda, Maryland: US Department of Health and Human Services. (*See* entry 321.)

348 **Zuckerman, A. J.** (Ed.). *A decade of viral hepatitis.* London: Elsevier/North-Holland, 1980. (512 pp.)

Methodology

349 **Palmer, W. J.** *Rapid and automated methods in microbiology and immunology: a bibliography 1976–1980.* London: Information Retrieval Ltd, 1981. (265 pp.) ISBN 0-904147-19-3.

350 **Voller, A.** *et al. The enzyme-linked immunosorbent assay (ELISA).* Guernsey: Flowline Publications, 1977. (48 pp.) ISBN 0-906146-00-3.

Plant Viruses

351 **Beale, H. P.** *Bibliography of plant viruses and index to research.* New York: Columbia University Press, 1976. (1500 pp.) ISBN 0-231-03763-5.

352 **Safferman, R. S.** and **Morris, M. E.** *Phycovirus bibliography.* Cincinnati, Ohio: Environmental Protection Agency, 1977. (17 pp.)

353 **Safferman, R. S.** and **Rohr, M. E.** *The practical directory for phycovirus literature 1979.* Cincinnati, Ohio: Environmental Protection Agency, 1980. (35 pp.)

Bibliographies Covering Books (*see* section 5.5)

354 *Books in print.* New York: Bowker, 1948–. Annual (fiche edition monthly or quarterly). ISSN 0068-0214.

355 *British books in print.* London: Whitaker, 1874–. Annual (fiche edition monthly). ISSN 0068-1350.

356 *British national bibliography.* London: British Library, 1950–. Weekly. ISSN 0007-1544.

357 *Index to Book Reviews in the Sciences.* Philadelphia: Institute for Scientific Information, 1980–. Monthly. ISSN 0196-447X.

358 *Index to Scientific Reviews.* Philadelphia: Institute for Scientific Information, 1975–. Quarterly and annual.

359 *MARCFICHE.* Washington, DC: Library of Congress, 1968–. Weekly.

360 *Subject guide to books in print.* New York: Bowker, 1957–. Annual. ISSN 0000-0159.

360A *Scientific and technical books and serials in print.* New York: Bowker, 1972–. Annual. ISSN 0000-054X.

360B *Medical books and serials in print.* New York: Bowker, 1972–. Annual. ISSN 0000-0574.

361 *World Health Organization publications: Catalogue, 1947–1973.* Geneva: World Health Organization, 1974. (206 pp.)

Directories of Conference Proceedings (*see* section 5.6)

362 *Conference Papers Index.* Riverdale, Maryland: Cambridge Scientific Abstracts, 1977–. Monthly. ISSN 0162-704X.

363 *Directory of Published Proceedings.* Series SEMT (Science, Engineering, Medicine and Technology). Harrison, New York: Interdok, 1965–. 10 p.a. ISSN 0012-3293.

364 *Index of Conference Proceedings Received by the BLL.* Wetherby, West Yorkshire, UK: British Library Lending Division, 1964–. Monthly. ISSN 0305-5183.

365 *Index to Scientific and Technical Proceedings.* Philadelphia: Institute for Scientific Information, 1978–. Monthly. ISSN 0149-8088.

366 *Proceedings in Print.* Arlington, Massachusetts: Proceedings in Print Inc., 1964–. Bi-monthly. ISSN 0032-9568.

Indexes to Dissertations (*see* section 5.7)

367 *American Doctoral Dissertations.* Ann Arbor, Michigan: University Microfilms International, 1934–. Annual.

368 *Abstracts of Theses.* London: Aslib, 1978–. Semi-annual. Microfiche only.

369 *British Reports, Translations and Theses.* Wetherby, West Yorkshire, UK: British Library Lending Division, 1971–. Monthly. ISSN 0144-7556.

370 *Comprehensive Dissertations Index.* Ann Arbor, Michigan: University Microfilms International, 1973–. Annual.

371 *Dissertation Abstracts International;* Section B, *The Sciences and Engineering.* Ann Arbor, Michigan: University Microfilms International, 1938–. Monthly. ISSN 0419-4217.

372 *Dissertation Abstracts International;* Section C, *European Abstracts.* Ann Arbor, Michigan: University Microfilms International, 1976–. Quarterly.

373 *Index to theses.* London: Aslib, 1950–. Semi-annual. ISSN 0073-6066.

374 *Masters Abstracts.* Ann Arbor, Michigan: University Microfilms International, 1962–. Quarterly. ISSN 0025-5106.

Classification Schemes (*see* section 5.8)

375 **Barnard, C.** *A classification for medical and veterinary libraries.* London: H. K. Lewis, 1955. (279 pp.)

376 *Bliss Bibliographic Classification.* 2nd edn. (Edited by J. Mills and V. Broughton.) London: Butterworths, 1980. ISBN 0-408-7082-X.

377 *Dewey Decimal Classification.* 19th edn. Albany, New York: Forest Press, 1979. 3 vols. (2692 pp.) ISBN 0-910608-23-7.

378 *Library of Congress Classification.* Class Q, *Science.* 6th edn. Washington, DC: Library of Congress, 1973. (415 pp.) ISBN 0-8444-0075-0.

379 *National Library of Medicine Classification.* 3rd edn. Bethesda, Maryland: National Library of Medicine, 1964. (286 pp.)

380 *Universal Decimal Classification: biological sciences.* 2nd edn. London: British Standards Institution, 1979.

MISCELLANEOUS PUBLICATIONS MENTIONED IN THE TEXT

381 *AGREP — Permanent Inventory of Agricultural Research Projects in the European Community.* Luxembourg: Commission of the European Communities, 1980. 2 vols: vol. 1 (841 pp.); vol. 2 (230 pp.).

382 *ASM News.* Washington, DC: American Society for Microbiology, 1913–. Monthly. ISSN 0044-7897.

383 *Bulletin d'Information/Newsletter.* Geneva: International Association of Biological Standardization, 1953–. Quarterly.

384 *Bulletin of the World Health Organization.* Geneva: World Health Organization, 1947–. Bi-monthly. ISSN 0042-9686.

385 *Indexing procedures for fifteen virus diseases of citrus trees* Indio, California: US Department of Agriculture, Handbook 333. 1968. (96 pp.)

386 *Lablore.* London: Wellcome Reagents, 1975–. Monthly.

387 *Nucleus.* Irvine, Scotland: Flow Laboratories, 1975–. Irregular.

388 *1982 Slide Catalog.* Washington, DC: American Society for Microbiology, 1981.

389 *Rabies Bulletin — Europe.* Tübingen: WHO Collaborating Centre for Rabies Surveillance and Research, 1975–. Quarterly.

390 *The Work of the WHO.* Geneva: World Health Organization, 1947–. Biennial. ISSN 0085-8285.

391 *WHO Chronicle.* Geneva: World Health Organization, 1947–. Monthly. ISSN 0042-9694.

392 *World Health Statistics Quarterly.* Geneva: World Health Organization, 1947–. Quarterly. ISSN 0043-8510.

Part III

DIRECTORY OF ORGANIZATIONS, CULTURE COLLECTIONS AND LIBRARIES

Directory of organizations, culture collections and libraries

Addresses of selected organizations	200
International (including commercial)	200
UK (Agricultural Research Council; Ministry of Agriculture, Fisheries and Food; Medical Research Council; Natural Environment Research Council; Public Health Laboratory Service; National Biological Standards Board; Others; Universities)	201
USA	205
Other countries (Australia; Canada; France; Federal Republic of Germany; India; Japan; USSR)	205
WHO Collaborating Centres (for Virus Reference and Research; Virus Collaborating Centres; for Arbovirus Reference and Research; for Reference and Research on Viral Hepatitis; for Reference and Research on Influenza; for Reference and Research on Smallpox; for Collection and Evaluation of Data on Comparative Virology; for Virus Reference and Research, Special Pathogens; for Reference and Research on Respiratory Viruses and Enteroviruses; on Food Virology; for Reference and Research on Rapid Laboratory Virus Diagnosis; for Reference and Research on Rotaviruses; for Reference and Research in Simian Viruses)	208
Major virus culture collections	215
Libraries	217
UK	217
USA and Canada	220
Other libraries	222

ADDRESSES OF SELECTED ORGANIZATIONS
International (including commercial)   (*see also* WHO Collaborating Centres, entries 479–529)

393 European Association against Virus Diseases
c/o Institute for Biomedical Sciences
University of Tampere
PO Box 607
SF-33101 Tampere 10
Finland

394 International Agency for Research on Cancer
16 Avenue Maréchal-Foch
60 Lyon (6e)
France

395 International Union of Microbiological Societies
31 Chemin Joseph-Aiguier
F-13274 Marseille Cédex 2
France

396 Commission of the European Communities
BP 1002
Luxembourg

397 International Association of Biological Standardization
Case Postale 229
CH-1211 Geneva 4
Switzerland

398 World Health Organization
Virus Diseases Unit
CH-1211 Geneva 27
Switzerland

399 Commonwealth Agricultural Bureaux
Farnham House
Farnham Royal
Slough
Buckinghamshire SL2 3BN
UK

400 Federation of European Microbiological Societies
c/o Meat Research Institute
Langford
Bristol BS18 7DX
UK

401 Smith Kline Corporation
Research Institute
The Frythe
Welwyn
Hertfordshire
UK

402 Wellcome Foundation Ltd
Wellcome Research Laboratories
Langley Court
Beckenham
Kent BR3 3BS
UK

403 International Organization of Citrus Virologists
455 Clinton Street
Indio
California 92201
USA

404 International Comparative Virology Organization
Institut für Medizinische Mikrobiologie, Infections- und Seuchenmedizin
Fachbereit Tiermedizin
Universität München
Veterinärstrasse
D-8000 München
Federal Republic of Germany

UK

Agricultural Research Council

405 Institute for Research on Animal Diseases
Compton
Near Newbury
Berkshire RG16 0NN

406 Animal Diseases Research Association
Moredun Institute
Gilmerton Road
Edinburgh ER17 7JH

407 Houghton Poultry Research Station
Houghton
Huntingdon
Cambridgeshire PE17 2DA

408 Glass House Crops Research Institute
Worthing Road
Littlehampton
West Sussex BN16 3PU

409 John Innes Institute
Colney Lane
Norwich NR4 7UK

410 Animal Virus Research Institute
Pirbright
Woking
Surrey GU24 0NF

Ministry of Agriculture, Fisheries and Food

411 Plant Pathology Laboratory
Hatching Green
Harpenden
Hertfordshire AL5 2BD

412 Veterinary Research Laboratories
Northern Ireland Department of Agriculture
Stormont, Storey Road
Belfast BT4 3SD

413 Central Veterinary Laboratory
New Haw
Weybridge
Surrey KT15 3NB

Medical Research Council

414 Laboratory of Molecular Biology
Hills Road
Cambridge CB2 2QH

415 Virology Unit
Institute of Virology
Church Street
Glasgow G11 5JR

416 Clinical Research Centre
Watford Road
Harrow
Middlesex HA1 3UJ

417 National Institute for Medical Research
Mill Hill
London NW7 1AA

418 Common Cold Unit
Harvard House
Coombe Road
Salisbury
Wiltshire

Natural Environment Research Council

419 Institute of Virology
5 South Parks Road
Oxford OX1 3UB

Public Health Laboratory Service

420 Central Public Health Laboratory
Colindale Avenue
London NW9 5HT

421 Communicable Disease Surveillance Centre
61 Colindale Avenue
London NW9 5EQ

422 Centre for Applied Microbiology and Research
Porton Down
Salisbury
Wiltshire SP4 0JG

National Biological Standards Board

423 National Institute for Biological Standards and Control
Holly Hill
Hampstead
London NW3 6RB

Others

424 Cancer Research Campaign
2 Carlton House Terrace
London SW1Y 5AR

425 Chester Beatty Research Institute
Fulham Road
London SW3 6BJ

426 Imperial Cancer Research Fund
Lincoln's Inn Fields
London WC2A 3PX

427 Institute of Cancer Research
Clifton Avenue
Belmont
Sutton
Surrey SH2 5PX

428 Society for General Microbiology
Harvest House
62 London Road
Reading RG1 5AS

Universities

429 Department of Microbiology and Immunobiology
The Queen's University of Belfast
Grosvenor Road
Belfast BT12 6BN

430 Division of Virology
Department of Pathology
University of Cambridge
Hills Road
Cambridge CB2 2QQ

431 University of Warwick
Coventry CV4 7AL

432 Heriot-Watt University
Edinburgh EH1 1HX

433 Institute of Virology
University of Glasgow
Church Street
Glasgow G11 5JR

434 Charing Cross Hospital Medical School
St Dunstan's Road
London W6 8RP

435 King's College Hospital Medical School
Denmark Hill
London SE5 8RX

436 Department of Medical Microbiology
London School of Hygiene and Tropical Medicine
London WC1E 7HT

437 Virology Department
St Bartholomew's Hospital
West Smithfield
London EC1A 7BE

438 St George's Hospital Medical School
Cranmer Terrace
Tooting
London SW17 0RE

439 Department of Virology
St Mary's Hospital Medical School
London W2 1PG

440 Department of Virology
St Thomas's Hospital Medical School
London SE1 7EH

441 Department of Bacteriology and Virology
University of Manchester Medical School
Oxford Road
Manchester M13 9PT

442 University of Reading
Reading
Berkshire RG6 2AH

DIRECTORY OF ORGANIZATIONS, CULTURE COLLECTIONS AND LIBRARIES   205

443 Department of Virology
University of Sheffield Medical School
Beech Hill Road
Sheffield S10 2RX

444 Sunderland Polytechnic
Chester Road
Sunderland SR2 3SD

445 Brunel University
Uxbridge
Middlesex UB8 3PH

USA

446 American Society for Microbiology
1913 I Street, NW,
Washington, DC 20001

447 Center for Disease Control
Atlanta
Georgia 30333

448 National Academy of Sciences
2101 Constitution Avenue
Washington, DC 20418

449 National Animal Disease Center
US Department of Agriculture
Ames, Iowa 50010

450 National Cancer Institute
National Institute of Health
Bethesda, Maryland 20205

451 National Institute of Allergy and Infectious Diseases
National Institute of Health
Bethesda, Maryland 20205

Other countries

Australia

452 Institute of Medical and Veterinary Science
Frome Road
Adelaide
South Australia 5000

453 The John Curtin School of Medical Research
Australian National University
Canberra City ACT 2601

454 CSIRO Division of Animal Health
Animal Health Research Laboratory
Private Bag No. 1
Parkville
Victoria 3052

455 National Biological Standards Laboratory
Private Bag No. 7
Parkville
Victoria 3052

Canada

456 Centre de Recherche en Virologie
Institut Armand-Frappier
531 boulevard des Prairies
Laval des Rapides, Quebec H7V 1B7

457 Department of Microbiology and Immunology
University of Montreal
CP 6128 Succursale 'A'
Montreal, Quebec H3C 3J7

458 Department of Microbiology and Parasitology
Faculty of Medicine
University of Toronto
Toronto, Ontario M5S 1A1

459 Faculty of Medicine
McGill University
845 Sherbrooke Street West
Montreal, Quebec H3A 2T5

France

460 Laboratoire de Génétique des Virus
CNRS
F-91190 Gif sur Yvette

461 Laboratoire des Virus
Institut Pasteur
20 boulevard Louis XIV
F-59012 Lille Cédex

462 Unité de Recherche sur les Infections Virales
Institut National de la Santé et de la Recherche Médicale
74 ave. Denfert-Rochereau
F-75014 Paris

Federal Republic of Germany

463 Institut für Virologie
Justus Liebig Universität
Frankfurterstrasse 107
D-6300 Giessen

464 Max von Pettenkofer-Institut
     Universität München
     Pettenkoferstrasse 9A
     D-8000 München 2

465 Institute for Virology and Immunobiology
     Versbacherstrasse 7
     D-8700 Würzburg

466 Max-Planck-Institut für Virusforschung
     Spemannstrasse 35
     Postfach 2109
     D-7400 Tübingen

India

467 Department of Virology
     Haffkine Institute
     Bombay 400012

468 Pasteur Institute of India
     Coonoor 643103 (Nilgiris)
     Taminadu State

469 Indian Society for Microbiology
     c/o Antibiotic Research Centre
     Pimpri
     Pune 411018

470 National Institute of Virology
     20-A Dr Ambedkor Road
     PO Box 11
     Pune 411001

Japan

471 Institute for Virus Research
     Kyoto University
     Kyoto

472 Department of Viral Infection
     Institute of Medical Science
     University of Tokyo
     PO Takanawa
     Tokyo

473 Virology Department
     Kikasato Institute
     5-9-1 Shirokane, Minato-Ku
     Tokyo 108

474 National Institute of Animal Health
1500 Josuihoncho
Kodaira
Tokyo

475 Society of Japanese Virologists
4-16 Yayoi
2-Chome
Bunkyo-Ku
Tokyo 113

USSR

476 All-Union Research Institute of Influenza
Ministry of Health
Prof. Popova Str. 15/17
Leningrad 197022

477 Institute of Experimental and Clinical Oncology
USSR Academy of Medical Sciences
Kashirskoye Shousse 6
Moscow 115478

478 Ivanovsky Institute of Virology
Ul. Gamaleya 16
Moscow 123098

WHO Collaborating Centres for Virus Reference and Research

479 Virus Research Laboratory
Faculty of Medicine
University of Ibadan
Ibadan
Nigeria

480 Laboratory of Infectious Diseases
National Institute of Allergy and Infectious Diseases
National Institutes of Health
Building 7, Room 301
Bethesda, Maryland 20014
USA

481 Department of Virology and Epidemiology
Baylor College of Medicine
Houston, Texas 77030
USA

482 Virus Research and Production Centre
Egyptian Organisation for Biological Products and Vaccine
Agouza
Cairo
Egypt

483 School of Public Health
Tehran University
PO Box 1310
Tehran
Iran

484 Institute of Hygiene and Epidemiology
Department of Epidemiology and Microbiology
Šrobárova 48
10042 Prague 10
Czechoslovakia

485 Laboratoire de Virologie
Université Claude-Bernard (Lyon I)
8 Avenue Rockefeller
F-69373 Lyon Cédex 3
France

486 Institut für Medizinische Mikrobiologie des Kantons St Gallen
Frohbergstrasse 3
CH-9000 St Gallen
Switzerland

487 Ivanovsky Institute of Virology
Ul. Gamaleya 16
Moscow 123098
USSR

488 MRC Common Cold Unit
Harvard Hospital
Coombe Road
Salisbury, Wiltshire
UK

489 Virus Laboratory
Fairfield Hospital
Yarra Bend Road
Fairfield, Victoria 3078
Australia

490 Department of Enteroviruses
National Institute of Health
4-7-1, Gakuen Musashimurayama
Tokyo 190-12
Japan

491 Department of Bacteriology
Faculty of Medicine
University of Singapore
Singapore 3

WHO Virus Collaborating Centres

492 Serviço de Virologia
     Instituto Adolfo Lutz
     Caixa Postal 7027
     São Paulo
     Brazil

493 Virus Diagnostic Services Division
     Bureau of Virology
     Laboratory Centre for Disease Control
     Department of National Health and Welfare
     Ottawa, Ontario K1A 0L2
     Canada

494 Department of Microbiology
     University of the West Indies
     Kingston 7
     Jamaica

495 Department of Virology
     National Institute of Hygiene
     Gyáli út 2-6
     1966 Budapest
     Hungary

496 'Stefan S. Nicolau' Institute of Virology
     285 Sos. Mihai Bravu
     74339 Bucharest
     Romania

497 Arbovirus Department
     Ivanovsky Institute of Virology
     Ul. Gamaleya 16
     Moscow 123098
     USSR

498 Laboratory of Mycoplasmas
     Gamaleya Institute of Epidemiology and Microbiology
     Ul. Gamaleya 18
     Moscow D-182
     USSR

499 Virus Reference Laboratory
     Central Public Health Laboratory
     Colindale Avenue
     London NW9.5HT
     UK

500 Virus Laboratory
     Andrija Stampar School of Public Health and
        Institute of Public Health of Croatia

Rockefellerova 12
YU-41000 Zagreb
Yugoslavia

WHO Collaborating Centres for Arbovirus Reference and Research

501 Institut Pasteur
36 avenue Pasteur
BP 220
Dakar
Senegal

502 East African Virus Research Institute
PO Box 49
Entebbe
Uganda

503 Vector-Borne Diseases Division
Center for Disease Control
Foothills Campus
PO Box 2087
Fort Collins, Colorado 80521
USA

504 Yale Arbovirus Research Unit
60 College Street
New Haven, Connecticut 06510
USA

505 Institute of Virology
Slovak Academy of Sciences
Mlynska dolina 1
Bratislava
Czechoslovakia

506 Unité d'Ecologie Virale
Institut Pasteur
F-75724 Paris Cédex 15
France

507 Queensland Institute of Medical Research
Bramston Terrace
Herston
Brisbane, Queensland 4006
Australia

508 Department of Virology and Rickettsiology
National Institute of Health
2-10-35 Kamiosaki, Shinagawa-Ku
Tokyo 141
Japan

WHO Collaborating Centres for Reference and Research on Viral Hepatitis

509 Center for Disease Control
Phoenix Laboratories Division
4402 N. 7th Street
Phoenix, Arizona 85014
USA

510 Department of Medical Microbiology
London School of Hygiene and Tropical Medicine
London WC1R 7HT
UK

511 Tokyo Metropolitan Institute of Medical Science
Honkomagome 3-18, Bunkyo-ku
Tokyo 113
Japan

WHO Collaborating Centres for Reference and Research on Influenza

512 Virology Division
Bureau of Laboratories
Center for Disease Control
Atlanta, Georgia 30333
USA

513 Virus Reference Laboratory
Central Public Health Laboratory
Colindale Avenue
London NW9 5HT
UK

514 Division of Virology
National Institute for Medical Research
Mill Hill
London NW7 1AA
UK

WHO Collaborating Centres for Reference and Research on Smallpox

515 Virology Division
Bureau of Laboratories
Center for Disease Control
Atlanta, Georgia 30333
USA

DIRECTORY OF ORGANIZATIONS, CULTURE COLLECTIONS AND LIBRARIES 213

516 Laboratoire National de la Santé
Ministère de la Santé
25 boulevard Saint Jacques
F-75014 Paris
France

517 Laboratory of Virology
Rijksinstituut voor de Volksgezondheid
PO Box 1
NL-3720 BA Bilthoven
Netherlands

518 Moscow Research Institute for Viral Preparations
Dubrovskaja ul. 15
Moscow 109088
USSR

519 Department of Virology
St Mary's Hospital Medical School
London W2 1PG
UK

520 Department of Enteroviruses
National Institute of Health
4-7-1, Gakuen Musashimurayama
Tokyo 190-12
Japan

WHO Collaborating Centre for Collection and Evaluation of Data on Comparative Virology

521 Institut für Medizinische Mikrobiologie, Infections- und Seuchenmedizin
Fachbereit Tiermedizin
Universität München, Veterinärstrasse,
D-8000 München
Federal Republic of Germany

WHO Collaborating Centres for Virus Reference and Research (Special Pathogens)

522 Virology Division
Bureau of Laboratories
Center for Disease Control
Atlanta, Georgia 30333
USA

523 Institute for Tropical Medicine
Nationalestraat 155
B-2000 Antwerpen
Belgium

524 Microbiological Research Establishment
Porton Down
Salisbury, Wiltshire
UK

WHO Collaborating Centre for Reference and Research on Respiratory Viruses and Enteroviruses

525 Virology Division
Bureau of Laboratories
Center for Disease Control
Atlanta, Georgia 30333
USA

WHO Collaborating Centre on Food Virology

526 Virology Section
Food Research Institute
University of Wisconsin
1925 Willow Drive
Madison, Wisconsin 53706
USA

WHO Collaborating Centre for Reference and Research on Rapid Laboratory Virus Diagnosis

527 Department of Virology
Royal Victoria Infirmary
Newcastle-upon-Tyne NE1 4LP
UK

WHO Collaborating Centre for Reference and Research on Rotaviruses

528 Regional Virus Laboratory
East Birmingham Hospital
Birmingham B9 5ST
UK

WHO Collaborating Centre for Reference and Research in Simian Viruses

529 Department of Microbiology and Infectious Diseases
Southwest Foundation for Research and Education
PO Box 28147
San Antonio, Texas 78284
USA

MAJOR VIRUS CULTURE COLLECTIONS

Australia

530 Department of Microbiology Culture Collection, University of Melbourne, Parkville, Victoria 3052
Holdings   Animal and bacterial viruses. Published in national catalogue.
Availability   Restricted

531 Virus Laboratory, Royal Children's Hospital, Flemington Road, Parkville, Victoria 3052
Holdings   Animal viruses. Published in national catalogue
Availability   Free of charge to other collections

532 Commonwealth Serum Laboratories Culture Collection, Poplar Street, Parkville, Victoria 3052
Holdings   Animal viruses. No catalogue
Availability   Free of charge to other collections

533 Victoria Plant Research Institute, Burnley Gardens, Swan Street, Burnley, Victoria 3121
Holdings   Plant viruses. No catalogue
Availability   Free to other collections

Canada

534 Microbiological Culture Collection, Public Health Laboratory, Ontario Department of health, PO Box 9000, Post Terminal A, Toronto 116, Ontario
Holdings   Animal viruses. Published in national catalogue
Availability   Free to hospitals, institutes and collections

535 Department of Plant Pathology, McDonald College, McGill University, Ste Anne de Bellevue, Quebec
Holdings   Plant viruses. Published in national catalogue
Availability   Personal favour

536 Insect Pathology Research Institute, Canadian Forestry Service, 1195 Queen Street E, PO Box 490, Sault Ste Marie, Ontario
Holdings   Insect viruses. No catalogue
Availability   Free to other collections

537 Canadian Communicable Disease Centre, Department of National Health and Welfare, Tunney's Pasture, Ottawa 3, Ontario
Holdings   Animal and bacterial viruses. Published in national catalogue
Availability   Free to other collections

538 Institut de Microbiologie et d'Hygiène, Université de Montréal, 531 boulevard des Prairies, Laval-des-Rapides, Québec

Holdings   Animal and bacterial viruses. Published in national catalogue
Availability   Free to other collections

Czechoslovakia

539   Czechoslovak National Collection of Type Cultures, Institute of Epidemiology and Microbiology, Srobárova 48, Prague 10
Holdings   Animal and bacterial viruses. Has own catalogue
Availability   At cost of postage and packaging but free to other collections

France

540   Collection du Service des Bacteriophages et des Bacteriocines, Institut Pasteur, 25 rue du Docteur Roux, Paris 15e
Holdings   Bacterial viruses
Availability   Unknown

541   Collection de Micro-organismes Associés aux Invertébrés, Station de Recherches Cytopathologiques INRA-CNRS, 30 Saint Christol-Montpellier, Gard
Holdings   Insect viruses
Availability   Free to other collections

German Democratic Republic

542   Fachgebiet Allgemeine Botanik und Pflanzenphysiologie, Ernst-Moritz-Arndt-Universität, Grimmerstrasse 88, 22 Greifswald, Bezirk Rostock
Holdings   Bacterial viruses. Publishes own catalogue
Availability   Free to other collections

Federal Republic of Germany

543   Robert Koch Institute, Federal Health Office, Nordufer 20, 1 Berlin 65, West Berlin
Holdings   Animal and bacterial viruses. No catalogue
Availability   At cost of packaging and posting

Japan

544   Culture Collection, National Institute of Animal Health, Japanese Federation of Culture Collections of Microorganisms, Josuihoncho 1500, Kodaira, Tokyo
Holdings   Animal viruses. Published in national catalogue
Availability   Free to other collections

545   Culture Collection Room, Research Institute for Microbial Diseases, Osaka University, Yamada-Kami, Suita, Osaka Prefecture
Holdings   Animal and bacterial viruses. Catalogue unknown
Availability   Unknown

UK

546 National Collection of Plant Pathogenic Bacteria, Plant Pathology Laboratory, Ministry of Agriculture, Fisheries and Food, Hatching Green, Harpenden, Hertfordshire
Holdings   Bacterial viruses
Availability   Not known

547 National Collection of Industrial Bacteria, Torry Research Station, PO Box 31, 135 Abbey Road, Aberdeen AB9 8DG
Holdings   Bacterial viruses. Has own catalogue
Availability   Fee payable but free to other collections

548 National Collection of Dairy Organisms, National Institute for Research in Dairying, Shinfield, Reading
Holdings   Bacterial viruses. Has own catalogue
Availability   Cost of postage and packaging, but free to other collections

549 Virology Department, Glasshouse Crops Research Institute, Littlehampton, West Sussex
Holdings   Plant viruses. No catalogue
Availability   Restricted

USA

550 American Type Culture Collection, 12301 Parklawn Drive, Rockville, Maryland 20852
Holdings   Animal, plant and bacterial viruses. Has own catalogue
Availability   Fee payable but free to other collections

551 Bioanalytical Culture Collection, Smith & French, 1500 Spring Garden Street, Philadelphia, Pennsylvania 19101
Holdings: Animal viruses. No catalogue
Availability   Free to other collections

LIBRARIES

UK

552 Animal Diseases Research Association, Moredun Institute, Gilmerton Road, Edinburgh
Specialism   Veterinary aspects of virology
Classification   Barnard
Holdings   7200 bound volumes; 1200 monographs; 150 current periodicals
Access   by arrangement
Publications   Annual list of papers by Association staff

218 DIRECTORY OF ORGANIZATIONS, CULTURE COLLECTIONS AND LIBRARIES

553 Animal Virus Research Institute, Pirbright, Woking, Surrey
 Specialism Animal virology, particularly foot-and-mouth disease and swine vesicular disease virus. Has a special collection of reprints on foot-and-mouth disease
 Classification Dewey
 Holdings 400 bound volumes; 100 current journals
 Access by arrangement
 Publications Information sheet on all recent papers on virus diseases of interest plus bibliographies on the same

554 Beecham Pharmaceutical Research Division, Clarendon Road, Worthing, West Sussex
 Specialism Chemotherapy of viral diseases
 Classification Universal Decimal Classification
 Holdings 5000 bound volumes; 300 monographs; 80 current periodicals
 Access by arrangement

555 Cambridge University, Department of Pathology, Kanthack Library, Tennis Court Road, Cambridge
 Specialism General virology, cancer
 Access Members of University only

556 Central Veterinary Laboratory, New Haw, Weybridge, Surrey
 Specialism Veterinary science, animal virology, Newcastle disease (also Library of Commonwealth — Bureau of Animal Health)
 Classification Barnard (modified)
 Holdings 36 000 bound volumes; 7000 monographs; 950 current periodicals; slides; reprints
 Access by arrangement
 Publications *Index Veterinarius, Veterinary Bulletin, Review Series*

557 Centre for Applied Microbiology and Research, Porton Down, Salisbury, Wiltshire
 Specialism Medical virology, including exotic viruses
 Holdings 3000 books; 140 current serials
 Access by arrangement
 Publications List of periodical holdings, list of recent accessions

558 Department of Health and Social Security, Alexander Fleming House, Elephant and Castle, London SE1 6BY
 Specialism Vaccination, smallpox
 Classification Bliss (modified)
 Holdings 200 000 bound volumes; 1300 current periodicals
 Access by arrangement
 Publications List of Departmental Publications, Index to Departmental Circulars, Bibliographies, Research and Development Report, and Handbook

559 Glasgow University Library, Hillhead Street, Glasgow G12 8QE
  Specialism Medical and general virology
  Classification Own, based on Library of Congress
  Holdings 70 000 bound volumes; 20 000 monographs; 1260 current periodicals
  Publications *Directory of Departmental Libraries*

560 Imperial Cancer Research Fund Laboratories, PO Box 123, Lincoln's Inn Fields, London WC2
  Specialism Cancer research, including oncogenic viruses
  Classification National Library of Medicine
  Holdings 10 000 bound volumes; 2000 monographs; 370 current periodicals
  Access no restrictions

561 Institute of Medical Laboratory Sciences, 12 Queen Anne Street, London W1M 0AU
  Specialism Clinical virology, immunology, theses submitted for fellowships
  Classification own
  Holdings 868 bound volumes; 500 monographs, 86 current periodicals
  Access by arrangement

562 John Innes Institute, Colney Lane, Norwich NR4 7UH
  Specialism Viral genetics, biochemistry and biophysics of plant viruses, electron microscopy
  Classification Universal Decimal Classification
  Holdings 40 000 bound volumes; 5000 monographs; 20 000 classified reprints; 250 current periodicals
  Access no restriction

563 Lilly Research Centre Ltd, Erl Wood Manor, Windlesham, Surrey GU20 6PH
  Specialism Veterinary and agricultural aspects of virology, chemotherapy
  Classification Universal Decimal Classification
  Holdings 6500 bound volumes; 2500 monographs; 200 current periodicals; tapes
  Access by post and telephone

564 London School of Hygiene and Tropical Medicine, Keppel Street, London WC1E 7HT
  Specialism Viral vaccination, history, exotic viruses, epidemiology, medical virology, smallpox
  Classification Barnard
  Holdings 63 000 bound volumes; 21 000 monographs; 1200 current periodicals; 1000 slides; tapes
  Access reference only

220 DIRECTORY OF ORGANIZATIONS, CULTURE COLLECTIONS AND LIBRARIES

Publications   Ross Institute Information and Advisory Service Bulletin, Tropical Diseases Bulletin

565 Medical Research Council, National Institute for Medical Research, The Ridgeway, Mill Hill, London NW7 1AA
Specialism   Medical virology
Classification   Barnard
Holdings   42 000 bound volumes; 9000 monographs; 700 current periodicals
Access   by post
Publications   Influenza Bibliography

566 Plant Pathology Laboratory, Hatching Green, Harpenden, Hertfordshire AL5 2BD
Specialism   Plant pathology and viruses
Holdings   5000 books; 3500 pamphlets; 550 current serials; 3000 photographs
Access   by arrangement

567 Royal Postgraduate Medical School, Wellcome Library, Hammersmith Hospital, Ducane Road, London W12 0HS
Specialism   General medical virology
Classification   Universal Decimal Classification
Holdings   22 000 current periodicals; 3000 monographs; 600 current periodicals; tapes; slides
Access   Reference only

568 Smith & Nephew Research Ltd, Gilston Park, Harlow, Essex
Specialism   Biochemical engineering aspects of virology, general virology
Classification   Universal Decimal Classification
Holdings   3000 bound volumes; 1000 monographs; 200 current periodicals
Access   Personnel

569 Wellcome Research Laboratories, Langley Court, Beckenham, Kent
Specialism   Medical virology, chemotherapy, immunology
Classification   Universal Decimal Classification
Holdings   25 000 bound volumes; 6000 monographs; 700 current periodicals; tapes; slides
Access   Personnel

USA and Canada

570 American Society for Microbiology, Archives Antonio Raimo, University of Maryland, Baltimore County 5401
Specialism   Monographs, theses and reprints
Holdings   3000 bound volumes

571 Canada Agriculture, Research Branch, Neatby Building, Rm 3032, Ottawa, Ontario
Specialism Animal and plant virology
Holdings 1100 volumes

572 Christ Hospital Institute of Medical Research, 2141 Auburn Avenue, Cincinatti, Ohio 45219
Specialism Medical virology
Holdings 16 000 volumes

573 Cold Spring Harbor Laboratory, Cold Spring Harbor, New York 11724
Specialism Viral genetics
Holdings 10 000 volumes; 15 000 monographs

574 Iowa State University, Ames, Iowa 50011
Specialism General virology

575 University of Massachusetts, Amherst Libraries, Amherst, Massachusetts 01003
Specialism General virology and immunological aspects
Holdings 70 000 bound volumes

576 Ohio Agricultural Research and Development Center, Madison Avenue, Wooster, Ohio 44691
Specialism Plant viruses, especially those of crops

577 Ontario Ministry of Health, Toronto, Ontario
Specialism Medical and applied virology
Holdings 4000 volumes

578 Pennsylvania State University, University Park, Pennsylvania 16802
Specialism Experimental virology
Holdings 108 000 volumes

579 Rutgers, The State University, Piscataway, Waksman Institute of Microbiology, PO Box 759, Piscataway, New Jersey 08854
Specialism Basic and applied virology
Holdings 16 000 volumes

580 Sunkist Growers Inc., Research Library, 760 Sunkist Street, Ontario, California
Specialism Viruses of citrus fruits
Holdings 1500 volumes

581 Warner-Lambert/Park-Davis Research Library, Detroit, Michigan 48232
Specialism Applied virology
Holdings 14 609 volumes

582 Yale Medical Library, 333 Cedar Street, New Haven, Connecticut 06510
Specialism   Medical virology
Holdings   425 000 volumes

Other Libraries

583 Bibliothecas de la Universidad de Buenos Aires, Azcuénaga 280, Faculty of Medicine and Faculty of Agriculture and Veterinary Medicine, Argentina
Specialism   General virology
Holdings   800 000 volumes

584 Australian National University Library, Canberra, ACT 2600, Australia
Specialism   General virology
Holdings   182 000 volumes

585 Commonwealth Scientific and Industrial Research Organization Libraries, 314 Albert Street, East Melbourne, Victoria 3002, Australia
Specialism   Plant and animal virology

586 University of Queensland Library, St Lucia, Brisbane, Queensland 4067, Australia
Specialism   General virology
Holdings   1 061 624 volumes

587 Czechoslovak Academy of Sciences, Institute of Virology Library, Mlynska Dolina 1, Bratislava, Czechoslovakia
Specialism   General virology
Holdings   12 000 volumes

588 Indian National Scientific Documentation Centre, Hillside Road, Delhi 110012, India
Specialism   General virology
Holdings   72 700 volumes; 4000 current periodicals; supplies information to scientists on request; compiles bibliographies

589 Keio University Medical Library and Information Center, 35 Shinano-machi, Shinjuku-ku, Japan
Specialism   Medical virology
Holdings   127 600 volumes and 2000 current periodicals

590 Ministry of Agriculture, Forestry and Fisheries Library, 2-1 Kasumigaseki, 1-Chome, Chiyoda-ku, Tokyo 100, Japan
Specialism   Animal and plant virology
Holdings   100 000 volumes

591 Universität Zürich, Institut für Medizinische Mikrobiologie, Bibliothek, Gloriastrasse 32, CH-8006 Zürich, Switzerland
Specialism   Medical virology

DIRECTORY OF ORGANIZATIONS, CULTURE COLLECTIONS AND LIBRARIES 223

592 Institute of Microbiology and Virology, Ul. Akademika Zabolotnogo 26, Kiev, Ukraine, USSR
Specialism General virology
Holdings 56 000 items

593 L'vov Research Institute of Epidemiology and Microbiology, Zelena 12, L'vov, Ukraine, USSR
Specialism General virology
Holdings 24 000 items

594 Universität Jena, Institut für Mikrobiologie, Bibliothek, Beuthenbergstrasse 11, Jena, German Democratic Republic
Specialism General virology
Holdings 3000 volumes

595 Universitätsbibliothek, Würzburg, Domerschulstrasse 16, D-8700 Würzburg, Federal Republic of Germany
Specialism Epidemiology and distribution of viral diseases
Holdings 5700 volumes

Index

Bold-face numerals indicate a reasonably substantial entry, or the most significant of several references to a topic. Authors' names are not indexed, nor are book titles except when mentioned in the text.

abstracting and indexing services **120–136**
  Biological Abstracts **121–124**, 134–135, 136
  comparative assessment **133–136**
  Excerpta Medica **124–125**, 134–135, 136, 141
  ICRDB Cancergrams 126
  Index Medicus **126–127**, 134–135, 136, 138
  Index Veterinarius 36, **127–128**, 134–135, 136
  International Abstracts of Biological Sciences **128**
  Review of Plant Pathology 36, 88, **129**, 141
  Science Citation Index **129–131**
  in subjects related to virology 133
  Veterinary Bulletin 36, **131**, 138
  Virology Abstracts 9
    abstracting and indexing 120–121, **131–132**
    comparative assessment 134–135
    conference proceedings 145
    dissertations 146
    listing journals 118
    ranking virology journals 74
    tracing and locating books 141
  Zentralblatt für Bakteriologie, Parasitenkunde, Infektionskrankheiten und Hygiene 3, **132–133**
Abstracts of Annual Meetings of the American Society for Microbiology 53, 66
Abstracts of Theses 146
Academy of Medical Sciences, USSR 62
Acta Virologica 73, **76**, 79, 139
Advances in Virus Research 8, 75, **88–89**
AGREP — Permanent inventory of agricultural research projects in the European Community 37
Agricultural Development and Advisory Service (ADAS) 43
Agricultural Research Council (ARC) (*see also* under individual laboratories, e.g. Animal Virus Research Institute) **41–43, 201–202**
  as funders of virus research 49
Albert Einstein College of Medicine 56
All-Union Influenza Research Institute, Leningrad 62
American Cancer Society 54, 55
American Doctoral Dissertations 145
American men and women of science: physical and biological sciences 113–114
American Society for Microbiology (ASM) (*see also ASM News*) 8–9, **53–54**
  education 54
  founding 6
  publications 53–54, 65–66
American Type Culture Collection (ATCC) 162, 163, 164
Animal Disease Occurrence 36–37
Animal Disease Research Association 41
Animal microbiology (Buxton and Fraser) 106
animal virology
  books on, *see* veterinary virology, books on
  history 3–4, 7–8
animal viruses, bibliographies 138, **191–192**
Animal Virus Research Institute, Pirbright 40, **41**, 162
Annales de l'Institut Pasteur 4

Annual Meetings of the American Society for Microbiology 65–66
Antimicrobial Agents and Chemotherapy 53, **77**
Antiviral Research **77**
Aphids as virus vectors (Harris and Maramorosch) 100
Applied and Environmental Microbiology 53
Arbovirus Research Program 59
Archiv für die Gesamte Virusforschung, see Archives of Virology
Archives of Virology 4, 7, 73, **77–78**
Argentina, library 222
Arthropod-borne Animal Disease Research Laboratory, Denver 52
ASM News 53
 guide to forthcoming conferences 64
 tracing and locating books 139
Assaad, F. 31
Assam Agricultural University 61
Association of Microbiologists of India 61
associations, professional
 UK 46–47
 USA 53–54
An atlas of insect and plant viruses (Maramorosch) 105
Australia
 culture collections **215**
 libraries **222**
 organizations **58–59, 205–206**, 209, 211
Australian National Animal Health Laboratory, Geelong 59
Australian National University 59

bacterial viruses, see bacteriophages
The bacteriophage (d'Hérelle) 5
The bacteriophage lambda (Hershey) 99
Bacteriophages (Adams) 99
Bacteriophages (Douglas) 99
bacteriophages
 bibliographies 138, 192
 books on **99, 179–180**
 history 5, 7, 9
Bakteriophagie (Raettig) 138
Barnard Classification Scheme 147, 148, **156–157**
Baylor College of Medicine 55
Beecham Pharmaceuticals 40
Belgium, organization 213
Beltsville Agriculture Research Center 52
Bergey's Manual of determinative bacteriology 11
bibliographies, subject **137ff., 191–193**

Bibliography of plant viruses and index to research (Beale) 139
Biochemical methods in cell culture and virology (Kuchler) 102
The biochemistry of viruses (Levy) 102
The biochemistry of viruses (Martin) 102
Biochimica et Biophysica Acta 75, **78**
Biogen Inc. 40
biographical dictionaries **112–114**
Biohazard control and containment in oncogenic virus research (Hellman) 167
Biological Abstracts **121–124**, 134–135, 136
Biological and biomedical resource literature (Kerker and Murphey) 116
Biological carcinogenesis research bibliography 138
Biologist, The 15, 17, 64
Biology data handbook (Alton and Dittmer) 110
The biology of animal viruses (Fenner et al.) 97
The biology of the interferon system (de Maeyer) 99
Bioscience 64
BIOSIS, see Biological Abstracts
book reviews 139–140
books **94–106, 178–188**
 on bacteriophages **99, 179–180**
 characteristics of **95–96**
 on chemotherapy **99, 180–181**
 on classification of viruses **105**, 186
 general (see also Comprehensive virology and Monographs in virology) **97–98, 178–179**
 on the history of virology **98–99, 179**
 on immunology **100, 181**
 on interferon **99, 180–181**
 on invertebrate virology **100, 181**
 on medical virology (see also Progress in Medical Virology and Virology Monographs) **100–101, 181–182**
 on methodology and diagnosis (see also Methods in Virology) **101–102, 182–183**
 on molecular virology **102, 183–184**
 monographic series **90–94**, 178
 on plant virology **102–103**, 105, **184**
 on slow viruses **103, 184–185**
 on specific virus groups **103–104, 185–186**
 on structure and morphology **104–105**, 186
 on taxonomy **105**, 186
 tracing and locating **139–143**, 193

on tumour virology **105–106, 186–187**
on veterinary virology (*see also Handbuch der Virusinfektionen Tieren*) **106, 187–188**
on viroids **103, 184–185**
on zoonoses 106, 187–188
Books in print 142
Brazil, organization 210
Bres, P. 31
British books in print 142
British Library Automated Information Services (BLAISE) 140, 144
British National Bibliography (BNB) 118, **140**
British Reports, Translations and Theses **146**
British Union Catalogue of Periodicals (BUCOP) 119
Brunel University, Uxbridge 49–50
Bulletin d'Information/Newsletter 34, 64
Bulletin of the World Health Organization 32
Bundesforschungsanstalt für Viruskrankheiten der Tiere, Tübingen 60

CAB ABSTRACTS 128
Caldwell, I. Y., 17
California Institute of Technology (Caltech) 7, 55
Canada
 culture collections **215–216**
 libraries **221**
 organizations **59, 206**, 210
cancer and viruses, *see* tumour virology
Cancer Research 64, 75, **78**, 79
Cancer Research Campaign 51
cancer research campaigns, as funders of virus research 49
Catalog of research reagents 1978–1980 52
Catalogue of cultures in the National Collection of Plant Pathogenic Bacteria 164
catalogues, publishers' 142
Cell 75, **80**
Cell biology (Altman and Katz) 110
Cell membranes and viral envelopes (Blough and Tiffany) 102
Cells and tissue culture (Paul) 102
Celltech Ltd 40
Cellular Immunology Unit, Oxford 44
Center for Disease Control, Atlanta 51
Central Public Health Laboratory, Colindale 44–45, 162
Central Veterinary Laboratory, Weybridge 43
Centre de Génétique des Virus, Gif-sur-Yvette 60

Centre for Applied Microbiology and Research (CAMR), Porton 44, 45, 162
Centre National de la Recherche Scientifique (CNRS) 60
Cetus 40
Charing Cross Hospital, London 47
chemotherapy, books on **99, 180–181**
Chemotherapy and control of influenza (Oxford and Williams) 99
Chemotherapy of herpes simplex virus infection (Oxford et al.) 99
Citrus Virus Diseases 36
classification
 books on, *see* taxonomy and classification, books on
 of viruses, *see* taxonomy of viruses
Classification and nomenclature of viruses (Matthews) 105
classification schemes, libraries, *see* library classification schemes
Clinical Research Centre, Harrow 44
Cold Spring Harbor Conferences on Cell Proliferation 66
Cold Spring Harbor Laboratory 7
Cold Spring Harbor Symposia on Quantitative Biology 66
Commission of the European Communities 36
Commonwealth Agricultural Bureaux **36**, 127–128, 131, 143
Commonwealth Bureau of Animal Health 36, 138
Commonwealth Mycological Institute 36, 129
Commonwealth Scientific and Industrial Research Organization (CSIRO) (*see also CSIRO Research Programs*) **58–59**, 162
Communicable Disease Surveillance Centre, Colindale 44, 45
Comparative and veterinary medicine: a guide to the resource literature (Kerker and Murphey) 116
Comparative diagnosis of viral diseases (Kurstak and Kurstak) 101
Comparative Immunology, Microbiology and Infectious Diseases 74, **80**
 reviews in 88
Comparative virology (Maramorosch and Kurstak) 37
Comprehensive Dissertation Index 145
COMPREHENSIVE DISSERTATION INDEX 145–146
Comprehensive virology (Fraenkel–Conrat

and Wagner) **90–91**
Conference Papers Index 143, 144
Conference Proceedings Index 144
conferences **63–69, 173–174**
  guides to **64–65, 173–174**
  major **65–69**
  proceedings **107–108, 173–174**
    directories of **193–194**
  role of **63–64**
  tracing and locating **143–145**
Contributions to the systematic plant virology (Hanson) 105
Cruelty to Animals Act 1876 165
CSIRO Research Programs 59
culture collections **161–164, 215–217**
  Australia **215**
  Canada **215–216**
  catalogues 164
  Czechoslovakia 216
  France 216
  German Democratic Republic 216
  Germany (Federal Republic) 216
  importance of 161–162
  Japan 216
  and patents 163
  UK **217**
  USA 217
Cumitechs 54
Current Advances in Plant Sciences 135
current awareness **134–135**, 191
Current Contents — Agriculture, Biology and Environmental Sciences 135
Current Contents — Life Sciences 135
Current serials received 118–119
Czechoslovakia
  culture collection 216
  organization 209

A decade of viral hepatitis (Zuckerman) 138
Decimal Classification scheme, *see* Dewey Decimal Classification
Dent, A. J. 17
Department of Defense, USA 52–53
Department of Health and Social Security, *see* Public Health Laboratory Service
Descriptions of Plant Viruses 36, **89**
Destructive Pests and Diseases of Plant Order 1965 167
Developments in antiviral therapy (Collier and Oxford) 99
Developments in Biological Standardization 66–67
Dewey Decimal Classification 147, 148, **149–152**

diagnosis, *see* methodology and diagnosis, books
Diagnostic methods in clinical virology (Grist et al.) 101
Diagnostic procedures for viral, rickettsial and chlamydial infections (Lennette and Schmidt) 101
DIALOG 157
  Biological Abstracts 123
  Books in print 142
  CAB Abstracts 128
  Conference Papers Index 144
  Dissertation Abstracts 145
  Science Citation Index 130
  tracing organizations 26
  tracing scientists 114
  tracing serials and annuals 118
  US research in progress 25
  Virology Abstracts 132
dictionaries
  biographical **112–114**
  subject **110–112**
A dictionary of biochemistry (Steneish) 112
A dictionary of biological acronyms (Dupayrat) 112
A dictionary of immunology (Herbot and Wilkinson) 112
A dictionary of microbial taxonomy (Cowan and Hill) 111
A dictionary of microbiology (Singleton and Sainsbury) 111
A dictionary of virology (Rowson et al.) 111
directories of organizations **23–26**
Directory of British associations 26
Directory of cancer research information resources 117
Directory of colleges and universities granting degrees in microbiology (USA) 24, 54
Directory of European scientific organizations 26
Directory of Published Proceedings 144
Directory of publishing 142
Diseases of Animals Acts 166
Diseases of poultry (Hofstad et al.) 106
Dissertation Abstracts International **145**
dissertations **108**
  indexes to **194**
  tracing and locating **145–146**
DNA tumor viruses (Tooze) 106
doctorates, *see* dissertations

East Malling Research Station, Maidstone 43
education 14–15, **49–50**, 54

Effects of interferon on cells, viruses and the immune system (Geraldes) 99
Egypt, organization 208
Eisai Company Ltd, Tokyo 61
An electron microscopic atlas of viruses (Williams and Fisher) 105
Eli Lily and Co. 40
Encyclopaedia Britannica 109
Encyclopedia of associations 26
encyclopedias and handbooks **108–110, 188**
The enzyme-linked immunosorbent assay (ELISA) (Voller et al.) 139
Epidemiology Unit, Cardiff 44
Essential immunology (Roitt) 100
Essentials of medical virology (Pumper and Yamashiroya) 101, 110
European Association against Virus Diseases **35**, 66
European Molecular Biology Laboratory, Heidelberg 44, 60
European Patent Office (EPO) 163
Excerpta Medica **124–125**, 134–135, 136, 141

FAO 9, 32, 37, 38, 52
FEBS Proceedings of Meetings 67
federal statutes 167
Federation of European Microbiology Societies (FEMS) **35**, 46, 67
Federal Republic of Germany, *see* Germany (Federal Republic)
Federation Proceedings 76, 108
FEMS 35, 64
FEMS Symposia 67
Fighting the unseen: the story of viruses (Reidman) 99
Filterable viruses (Rivers) 5
Finland, organization 200
Flow laboratories 41, 42
Food and Agriculture Organization, *see* FAO
Food and Drug Administration 51, 52
Foot and Mouth Disease Bulletin 40
Forthcoming International Scientific and Technical Conferences 64, 65
France
  culture collections 216
  organizations **60**, 200, 206, 210
Frederick Cancer Research Center 56
fungal viruses, books on, *see* plant virology, books on

Gamaleya Institute of Epidemiology and Microbiology 62

general books on virology (*see also Comprehensive virology* and *Monographs in virology*) **97–98, 178–179**
General virology (Luria et al.) 97
The genetics of bacteria and their viruses (Hayes) 102
German Democratic Republic
  culture collection 223
  library 223
Germany (Federal Republic)
  culture collection 216
  library 223
  organizations **60**, 201, **206–207**, 213
Gibco Europe 41, 42
Glossary of genetics and cytogenetics (Rieger et al.) 112
Glossary of molecular biology (Evans) 112
Godber Report 165
Graduate studies (UCCA) 24
GRANTS 26
Grants Information System (GIS) 26
Grants register 26
Guide to the literature of the life sciences (Smith and Reid) 117
guides to the literature **115–117, 189–190**
Gustav Stern Symposia, *see Perspectives in Virology*
Guy's Hospital, London 44

Haffkine Institute, Bombay 60
Hagan and Bruner's Infectious diseases of domestic animals (Gillespie and Timoney) 106
Handbook of microbiology (Laskin and Lechevalier) 109–110
Handbook of plant virus infections (Kurstak) 103
handbooks, *see* encyclopedias and handbooks
Handbuch der Virusinfektionen Tieren (Rohrer) **91**
Harvard Medical School 56
Harvey Lectures 67
Heriot–Watt University 50
The herpesviruses (Kaplan) 103
history and development of virology **3–11**
  books on **98–99, 179**
  milestones in 4
Hoffman–La Roche 40
Houghton Poultry Research Station 43
How to apply for admission to a university (UCCA) 24
Howie Report 165–166

230 INDEX

human virology, *see* medical virology, books on
human viruses, bibliographies **138–139**, 192
Hungary, organization 210

ICN–UCLA Symposia on Molecular and Cellular Biology 68
ICRDB Cancergrams 126
immunology, books on **100, 181**
Imperial Cancer Research Fund 51
Index Medicus **126–127**, 134–135, 136, 138
Index of Conference Proceedings Received by the BLL **143**
Index to Book Reviews in the Sciences 140
Index to Scientific and Technical Proceedings 143, 144
Index to Theses 146
Index Veterinarius 36, **127–128**, 134–135, 136
Indexing procedures for fifteen virus diseases of citrus trees 36
India
  library 222
  organizations **61–62, 207**
Indian Council for Medical Research 60
Indian Journal of Microbiology 61
Industrial research in the United Kingdom 25
Infection and Immunity 53, 74, 79, **80–81**
Influenza: a Royal Society discussion (Tyrell and Pereira) 104
The influenza viruses and influenza (Kilbourne) 104
Information sources in agriculture and food science (Lilley) 117
Insect virology (Smith) 100
Institut Armand-Frappier, Quebec 59
Institut de Recherche en Biologie Moléculaire, Paris 60
Institut de Recherches Scientifiques sur le Cancer, Villejuif 60
Institut für Medizinische Mikrobiologie, Infektions- und Seuchenmedizin, Munich 37, 60
Institut für Virologie, Giessen 60
Institute of Biological Control 36
Institute of Cancer Research 51
Institute of Medical and Veterinary Research, Adelaide 59
Institute of Microbiology and All-Union Microbiology Society, USSR 61–62
Institute of Ophthalmology, University of London 47
Institute of Poliomyelitis and Viral Encephalitis, USSR 62
Institute of Terrestrial Ecology 45
Institute of Virology, Glasgow, *see* University of Glasgow
Institute of Virology, Oxford 45
Interferon (Gresser) 99
interferon, books on, *see* chemotherapy, books on
International Abstracts of Biological Sciences **128**
International Agency for Research on Cancer (IARC) **37**, 44
International Association of Biological Standardization (IABS) 8, **34–35**, 66–67
International Association of Microbiological Societies (IAMS), *see* International Union of Microbiological Societies
International Committee on Taxonomy of Viruses 12, **34**
International Comparative Virology Organization (ICVO) 9, **37–39**, 68
International Conference on Comparative Virology 9, 39, **68**
International Conference on Vaccines against Viral and Rickettsial Diseases of Man 9
International Congress for Microbiology 6
International Congress for Virology 68–69
International Journal of Cancer 79, **81**
International Organization of Citrus Virologists **35–36**
International Union of Microbiological Societies (IUMS) 9, 14, **33–34**, 46, 117
  conferences 68–69
  standardization, *see* International Association of Biological Standardization
  taxonomy, *see* International Committee on Taxonomy of Viruses.
  Virology Division 34
International Virology series 9, **68–69**
Intervirology 34, 73, 77, 79, **81**
  conference notices in 64
  reviews in 88
Introduction to modern virology (Primrose and Dimmock) 98
An introduction to the history of virology (Waterson and Wilkinson) 98
  biographies in, 112, 113
Introduction to virology (Smith and Ritchie) 98

invertebrate virology
  books on **100**, **181**
  history 10
Iran, organization 209
IRL Life Sciences Collection 132
Irregular serials and annuals 118
Ivanovsky Institute of Virology, Moscow 62

Jamaica, organization 210
Japan
  culture collections 216
  libraries 222
  organizations **61**, **207–208**, 209, 211, 212, 213
John Curtin School of Medical Research 59
John Innes Institute, Norwich 43
Johns Hopkins University 56
Journal of Bacteriology 53, 79, **81**
Journal of Biological Chemistry 75, 77, **81**
Journal of Biological Standardization 34
Journal of Clinical Microbiology 53, 74, **81–82**
Journal of Experimental Medicine 5, 75, **82**
Journal of General Microbiology 35, 47, 75, 79, **82**
Journal of General Virology 35, 47, 73, 77, 78, 79, **82**
  reviews in 88
Journal of Immunology 74, 78, 79, **82**
Journal of Invertebrate Pathology 74, 80, **82**
Journal of Medical Virology 73, 80, **82**
Journal of Molecular Biology 75, 78, **83**
Journal of Virological Methods 73, **83**
Journal of Virology 9, 53, 73, 75, 77, 78, 79, **83**
Journal of Wildlife Diseases 74, **83–84**
Journal of the National Cancer Institute 74, 78, 79, **83**
journals **72–87**, **174–177**
  articles, tracing and locating **120–134**
  characteristics of virology journals **72–76**
  general characteristics **72**
  growth of, in virology 8
  secondary **120ff.**, **190–191**
  tracing and locating **117–120**, **190**

Keyword Index to Serial Titles 119
King's College Hospital, London 47
Kitasato Archives of Experimental Medicine 61
Kitasato University 61

Lablore 40
Laboratoire Central des Recherches Vétérinaires, Maison Affore 60
Laboratory of Molecular Biology, Cambridge 44
Lancet, The 5, 64, 75, 77, 79, **84**
Lasswade Laboratory, Scotland 43
legislation and laboratory safety **165–168**
  UK **165–167**
  USA **167**
libraries **217–223**
  Argentina 222
  Australia **222**
  Canada **221**
  Czechoslovakia 222
  German Democratic Republic 223
  Germany (Federal Republic) 223
  India 222
  Japan 222
  Switzerland 222
  UK **217–220**
  USA **220–222**
  USSR 223
library classification schemes **147–157**, **194**
  Barnard 147, 148, **156–157**
  comparison 148
  'Dewey' 147, 148, **149–152**
  Library of Congress 147, 148, **154–155**
  National Library of Medicine **157**, **158**
  types 149
  Universal Decimal Classification **152–154**
Library of Congress Classification Scheme (LC) 147, 148, **154–155**
London School of Hygiene and Tropical Medicine 44, 47, 48
Luxembourg, organization 200

McGill University 59
McGraw–Hill encyclopedia of science and technology 109
MARCFICHE **141**
Massachusetts Institute of Technology 56
Masters Abstracts 145
Max-Planck-Gesellschaft zur Förderung der Wissenschaften EV (Max Planck Society for the Advancement of Science) 5, 60
Max-Planck-Institut für Virusforschung, Tübingen 60

Medical books and serials in print 142
Medical microbiology (Duguid *et al.*) 100
Medical Research Council (MRC) (*see also* individual laboratories, e.g. National Institute for Medical Research) **43–44, 202**
    as a funder of virus research 49
Medical virology (Fenner and White) 101
medical virology, books on (*see also Progress in Medical Virology* and *Virology Monographs*) **100–101, 181–182**
Medicines Act 1968   166
MEDLINE, *see Index Medicus*
Merck Frosst Laboratories 40
methodology
    bibliographies **139**, 193
    and diagnosis, books on (*see also Methods in Virology*) **101–102, 182**
Methods in Virology **91–92**
Michigan State University 57
Microbiology (Davis) 100
Microbiology (series) 53, 66, **108**
Microbiology and Immunology 61
Microbiology Letters 35
Microbiology Reviews 53
Microbios 74, 79, **84**
Ministry of Agriculture, Fisheries and Food (MAFF) (*see also* individual laboratories, e.g. Central Veterinary Laboratory) **43**, 166, **202**
Molecular and Cellular Biology 53
Molecular and General Genetics 75, 77
The molecular biology of animal viruses (Prosad-Nayak) 102
Molecular biology of bacterial viruses (Stent) 99
The molecular biology of picornaviruses (Perez-Bercoff) 104
Molecular biology of RNA tumor viruses (Stephenson) 106
Molecular virology (Knight) 102
Molecular virology (Pennington) 102
molecular virology, books on **102, 183–184**
monographs, *see* books
Monographs in Virology **92**
Moredun Institute, Edinburgh 41
morphology, books on, *see* structure and morphology, books on
Moscow Institute of Virus Preparations 62
Munich Symposia on Microbiology 39, 68

National Academy of Sciences (USA) **53**
National Animal Disease Center, Ames 52
National Biological Standards Laboratory, Canberra 59
National Cancer Institute 52
National Institute for Allergy and Infectious Diseases — catalogue of research reagents 1978–1979   164
National Institute for Biological Standards and Control (NIBSC) **45–46**
National Institute for Medical Research, Hampstead 5, 43–44, 46
National Institute of Agricultural Sciences, Tokyo 61
National Institute of Allergy and Infectious Diseases (NIAID) 52, 162
National Institute of Animal Health, Tokyo 61
National Institute of General Medical Sciences, USA 52
National Institute of Health, Tokyo 61
National Institutes of Health, USA **51–52**
National Institute of Neurological and Communicative Disorders and Stroke 52
National Library of Medicine bibliographies 138
National Library of Medicine classification scheme **157, 158**
National Vegetable Research Station, Wellsbourne 43
Natural Environment Research Council (NERC) **45**, 202
The natural history of rabies (Baer) 104
Nature 18, 64, 73, 75, 77, 78, 79, **84**
    reviews in 88
Naturwissenschaften 6
Netherlands, organization 213
Neurology 80, **84**
New England Journal of Medicine 64, 75, 79, **84**
New Research Centers 25
New Scientist 18, 64
New Serial Titles 118, 119
New York University School of Medicine 57
NIAH Quarterly 61
Nigeria, organization 208
Nihon Gakujutsu Kaigi (Science Council of Japan) 61
Nippon Shokubutsu–Byori Gakkai (Phytopathological Society of Japan) 61
Nippon Virusu Gakkai (Society of

Japanese Virologists) 61
nomenclature of viruses **11–12**
Non-arthropod-borne togaviruses (Horzinek) 104
North-West European Microbiological Group, *see* Federation of European Microbiology Societies
Nucleus 41

OCLC 120, 143
Oncogenic viruses (Gross) 105
on-line computer services (*see also* individual databases, e.g. DIALOG) 157
ORBIT 123, 130, 157
organizations **21–62, 199–214**
   Australia **58–59, 205–206,** 209, 211
   Belgium 213
   Brazil 210
   Canada **59, 206,** 210
   commercial **39–41,** 42, 200, 201
   Czechoslovakia 209
   directories of **172–173**
   Egypt 208
   finding out about **23–26**
   Finland 200
   France **60,** 200, 206, 210
   Germany (Federal Republic) **60,** 201, **206–207,** 213
   Hungary 210
   India **60–61, 207**
   as information sources 23
   international **28–39, 200–201, 208–214**
   Iran 209
   Jamaica 210
   Japan **61, 207–208,** 209, 211, 212, 213
   Luxembourg 200
   Netherlands 213
   Nigeria 208
   as publishers of information 23
   Romania 210
   Senegal 211
   Singapore 209
   Switzerland 200, 209
   types **22–23**
   Uganda 211
   UK **41–51,** 200, **201–205,** 210, 212, 213, 214
   USA **51–58,** 200, **205,** 208, 211, 212, 213, 214
   USSR **61–62, 208,** 209, 210, 213

Pasteur Institute, Canoor 61
patents, *see* culture collections

Pennsylvania State University 57
periodicals, *see* journals
Periodicals relevant to microbiology and immunology 117
Perspectives in Virology 8, **92–93**
Phage and the origins of molecular biology (Cairns *et al.*) 98
phages, *see* bacteriophages
Phycovirus bibliography (Safferman and Morris) 139
Phytopathology 64, 74, 80, **85**
Plant Disease 85
Plant Health Act and Orders (1967) 167
Plant Pathology 43
Plant Pathology Laboratory, Harpenden 43
Plant virology (Matthews) 103
plant virology (including fungal viruses) 105
   books on **102–103, 184**
   history 3, 6, 9–10
Plant virology: the principles (Gibbs and Harrison) 103
Plant viruses (Smith) 103
plant viruses 89
   bibliographies 139, 193
Plum Island Animal Disease Center, New York 52
The practical directory for phycovirus literature (Safferman and Rohr) 139
Principles of animal virology (Joklik) 98
Proceedings in Print 145
Proceedings of the National Academy of Sciences 53, 74, 77, 78, 79, **85,** 108
Proceedings of the Royal Society of London 108
Proceedings of the Society for Experimental Biology and Medicine 74, **85,** 108
Progress in Medical Virology 8, **93**
Public Health Laboratory Service (*see also* individual laboratories, e.g. Centre for Applied Microbiology and Research) **44–45, 203**

Queen's University, Belfast 48, 50

Rabies: the facts (Kaplan *et al.*) 104
Rabies Acts and Orders, various 166
Rapid and automated methods in microbiology and immunology: a bibliography 1976–1980 (Palmer) 139
Recent advances in clinical virology (Waterson) 101
reference works **108–114**

Regional Poultry Research Laboratory, East Lansing 52
Replication of mammalian parvoviruses (Ward and Tattersal) 104
Research center directory 25
Research in British universities polytechnics and colleges 25
Research Reagents Program 52
Review of Plant Pathology 36, 88, **129**, 141
review series **87–94, 177–178**
   characteristics **87–88**
   location 88
   subjects related to virology 94
Revue Roumaine de Médecine — Virologie 73, **86**
Rhabdoviruses (Bishop) 104
Ribavirin — a broad spectrum antiviral agent (Smith and Kirkpatrick) 99
RNA phages (Zinder) 99
RNA tumor viruses (Stephenson) 106
Rockefeller University 57
Romania, organization 210
Rothamsted Experimental Station, Harpenden 43
Royal Free Hospital, London 44
Royal Postgraduate Medical School, London 44

safety, *see* legislation and laboratory safety
St Bartholomew's Hospital, London 47
St George's Hospital, London 47
St Louis University Medical Center 57
St Mary's Hospital, London 47
St Thomas's Hospital, London 47
Salk Institute, San Diego 57
Science 64
Science Citation Index (SCI) **129–131**
Scientific and technical books and serials in print 142
Scottish Horticultural Research Institute, Dundee 43
Searle Research and Development Ltd 40
Selected papers in tumor virology (Tooze) 106
Selected papers on virology (Hahon) 98
Senegal, organization 211
Serials in the British Library 119
series, *see* review series
Shionogi Research Laboratories, Shiga 61
Singapore, organization 209
Single-stranded DNA phages (Denhardt) 99
Slide Catalog 54

Slow transmissible diseases of the nervous system (Prusiner and Hadlow) 103
Slow virus diseases of animals and man (Kimberlin) 103
Slow viruses (Adams and Bell) 103
slow viruses and viroids, books on **103, 184–185**
Smith Kline Corporation 40
Smithsonian Science Information Exchange (SSIE) 25
Society for General Microbiology (SGM) 8, 9, 35, 37, **46–47**
   conference proceedings 108
   conferences 46, 49
   publications 47
Society for General Microbiology Quarterly 47, 72
   abstracts of conferences 69
   book reviews 139
   guide to forthcoming conferences 64
Society for General Microbiology Symposia 69
Society of American Bacteriologists, *see* American Society for Microbiology
Soviet Progress in Virology, see Voprosy Virusologii
Specific treatment of virus diseases (Bauer) 99
specific virus groups, books on **103–104**
Structure and function of viruses (Horne) 105
structure and morphology, books on **104–105, 186**
subject dictionaries **110–112**
Subject guide to books in print 142
Sunderland Polytechnic 48
Switzerland
   library 222
   organizations 200, 209
symposia, *see* conferences
Symposia of the European Association against Virus Diseases 66
Symposium on fungal viruses (Molitoris et al.) 103

Tamil Nadu Agricultural University 61
taxonomy and classification, books on **105**, 186
taxonomy of viruses 11–12, 13
textbooks, *see* books
theses, *see* dissertations
The togaviruses: biology, structure and replication (Schlesinger) 104
Trade associations and professional bodies of the UK 26
treatises, *see* books

tumour virology
  books on **105–106, 186–187**
  history 9

Uganda, organization 211
UK
  culture collections **217**
  libraries **217–220**
  organizations **41–51**, 200, **201–205**, 210, 212, 213, 214
Ulrich's International periodical directory **118**, 142
Ulrich's Quarterly 118
Ultrastructure of animal viruses and bacteriophages — an atlas (Dalton and Haguenau) 105
Unilever Research 40
Unit of Invertebrate Virology, Oxford, *see* Institute of Virology, Oxford
United Kingdom Federation for Culture Collections Newsletter 161
Universal Decimal Classification (UDC) **152–154**
universities and colleges
  in the UK **203–204**
    research **47–49**
    research funding 49
    study **49–50**
  in the USA **54–58**
    research funding 54, 55
University of Alabama 54–55
University of Birmingham, UK 48, 50
University of Bristol 48, 50
University of California 55
University of Cambridge 5, 48
University of Chicago 56
University of Dundee 48, 50
University of Edinburgh 48, 50
University of Florida 56
University of Glasgow **48**, 50
University of Illinois 56
University of Iowa 56
University of Leeds 50
University of Liverpool 48
University of London **47**
University of Manchester 5, 48, 49, 50
University of Maryland 56
University of Minnesota 57
University of Missouri 57
University of Newcastle 48, 50
University of Oxford 48, 50
University of Queensland 59
University of Reading 48, 50
University of Rochester 57

University of Sheffield 48, 50
University of Southern California 57
University of Surrey 48, 50
University of Texas 58
University of Toronto 59
University of Utah Medical School 58
University of Warwick 48, 49
University of Wisconsin 58
University of Wisconsin — Madison 58
Upjohn Company 40
USDA Emergency Programs Foreign Animal Disease Data Bank, bibliographies 138
US Department of Agriculture **52**, 154, 167
US Department of Health and Human Services, *see* Center for Disease Control
US Patent Office 163
USA
  culture collections 217
  libraries **220–222**
  organizations **51–58**, 200, **205**, 208, 211, 212, 213
The use of biological literature (Bottle and Wyatt) 117
The use of medical literature (Morton) 116
USSR
  libraries 223
  organizations **61–62, 208**, 209, 210, 213

Vaccine preparation techniques (Duffy) 100
Van Nostrand's scientific encyclopedia 109
Vanderbilt University, Nashville 58
Veterinary Bulletin 36, **131**, 138
Veterinary encyclopaedia 109
Veterinary Record 64, 74, 80, **86**
Veterinary Research Laboratories, Belfast 43
Veterinary virology (Mohanty and Dutta) 106
veterinary virology, books on (*see also Handbuch der Virusinfektionen Tieren*) **106, 187–188**
Viral immunodiagnosis (Kurstak and Morisset) 100
Viral immunology and immunopathology (Notkins) 100
Viral infections of humans: epidemiology and control (Evan) 101
viroids, books on, *see* slow viruses
Viroids and viroid diseases (Diener) 103

virologists
  employment 15–18
  information needs of 18–20
Virology 8, 73, 75, 77, 78, 79, **86**
virology
  a definition 11–12
  as a profession **14–18**
  relationship to other disciplines 12, 14
Virology Abstracts 9
  abstracting and indexing 120–121, **131–132**
  comparative assessment 134–135, 136
  conference proceedings 145
  dissertations 146
  listing journals 118
  ranking virology journals 74
  tracing and locating books 141
Virology Literature 121
Virology Monographs **93–94**
Virology NIAID Task Force 52
Virus, 61, 73, **86**
Virus Cancer Program 52
The virus: a history of the concept (Hughes) 98
Virus hunters: the lives and triumphs of great modern medical pioneers (Williams) 99
Virus–insect relationships (Smith) 100
virus morphology, books on, **104–105**
Virus receptors 102
Virus Research Centre, Poona 60
Virus structure (Horne) 104–105
Viruses and immunity (Koprowski and Koprowski) 100
Viruses and invertebrates (Gibbs) 100
Viruses of vertebrates (Andrewes et al.) 98, 110
Virusy i Virusnye Zabolevaniya 73, **86–87**
Voprosy Virusologii (Soviet Progress in Virology) 73, 79, **87**

Walter Reed Army Medical Center, Washington 52–53
Wellcome Foundation 39–40
Wellcome Reagents 40, 41, 42
Wellcome Trust 40, 49
WHO, *see* World Health Organization
Who's Who in British Scientists 1980/81 114
Wistar Institute of Anatomy and Biology, Philadelphia 58
Wolverhampton Polytechnic 50
The Work of the WHO 32
World directory of collections of cultures of micro-organisms (Martin and Skerman) 162

World Federation for Culture Collections 162
World guide to scientific associations 26
World Health Organization (WHO) 8, 9, 22, 24, 27, **28–33**, 162
  Africa, special studies 31
  Centres for Virus Reference and Research 29
  Collaborating Centres 29, **208–214**
  Collaborating Centre for Collection and Evaluation of Data on Comparative Virology 37
  Collaborating Centre on viral hepatitis 47
  founding 4, 7
  function and organization **28–29**
  as funder of culture collections 162
  as funder of virus research 49
  Monograph series 32
  reagents programme 29, 31
  referral services 29
  research 29
  structure of organization 30
  Technical Report Series 32–33
  virus information systems **31–32**
WHO Chronicle 32
World Health Organization publications catalogue 33
World Health Statistics 33
World Influenza Centres 7
World Intellectual Property Organization (WIPO) 163
World list of virus laboratories 22, 24, 29
World Meetings: outside the United States and Canada 65
World Meetings: United States and Canada 65
World of learning 26
Wye College, University of London 48
Wyeth Laboratories 40

Yale University, New Haven 5, 58
Yearbook of international organizations 26

Zentralblatt für Bakteriologie, Mikrobiologie und Hygiene 64, 139
Zentralblatt für Bakteriologie, Parasitenkunde, Infektionskrankheiten und Hygiene 3, **132–133**
zoonoses, books on, *see* veterinary virology, books on

RAYMOND H. FOGLER LIBRARY
DATE DUE

SUBJECT TO
WEEKS